National Conference on Structural and Geotechnical Engineering 2022

IOP Conference Series: Materials Science and Engineering
Volume 1273

Jamshedpur, India
11 – 12 November 2022

ISBN: 978-1-7138-6766-1
ISSN: 1757-8981

Printed from e-media with permission by:

Curran Associates, Inc.
57 Morehouse Lane
Red Hook, NY 12571

Some format issues inherent in the e-media version may also appear in this print version.

Institute of Physics
Dirac House, Temple Back
Bristol BS1 6BE UK

Phone: 44 1 17 929 7481
Fax: 44 1 17 920 0979

techtracking@iop.org

Additional copies of this publication are available from:

Curran Associates, Inc.
57 Morehouse Lane
Red Hook, NY 12571 USA
Phone: 845-758-0400
Fax: 845-758-2633
Email: curran@proceedings.com
Web: www.proceedings.com

TABLE OF CONTENTS

IOP Conf. Series: Materials Science and Engineering 1273 (2023) 011001 doi:10.1088/1757-899X/1273/1/011001

National Conference on
STRUCTURAL AND GEOTECHNICAL ENGINEERING (NCSGE-2022)

11th and 12th November 2022

INTRODUCTION

Every structure requires foundation for its stability. A simple failure of foundation may lead to the failure of a structure even though all codal provisions are met in the analysis and design of super structure. A prompt exchange of ideas and innovations is essential between structural and geotechnical engineers to build a safe structure. Hence a single platform to exchange the knowledge between the experts of structural and geotechnical engineers is materialized in The first National Conference on **"Structural and Geotechnical Engineering (NCSGE-2022)"** . The Two Day National is organized by the Department of Civil Engineering, National Institute of Technology Jamshedpur, on November11-12, 2022. The main objective of this conference is to discuss the problems, solutions or recommendations in the fields of structural and geotechnical engineering. The experts from academic institutions, industries, research organizations, practicing engineers and students are given prime opportunity to share and discuss all the real challenges and issues in various fields like seismic studies, sustainable new materials, thermal studies, properties of black cotton soils, retaining walls and lateral load capacity etc. The main focus was given to interchange the ideas and novelty from structural engineering to geotechnical engineering and vice versa.

ABOUT THE CONFERENCE

The conference was inaugurated on 11th November 2022 with two keynote lectures by Prof. N. K. Samadhiya, from Indian Institute of Technology Roorkee and Prof. Karunesh Kumar Shukla, Director, National Institute of Technology Jamshedpur. Prof. N. K. Samadhiya presented online on "Analysis of shortcrete lining and estimation of support pressure of Sardar Sarovar underground powerhouse cavern". The analysis of tunnel cavern was demonstrated and stages of construction was discussed.
Prof. Karusnesh Kumar Shulka explained the concept of geometric nonlinearity and demonstrated the solution techniques of nonlinear equation of motion of continuous systems.
Prof. A. Murali Krishna from Indian Institute of Technology Tirupati greeted the organizing team for the successful organization of the conference. Totally fifteen research papers were presented in the field of structural engineering and twenty three research papers were discussed in the field of geotechnical engineering. All the research articles were discussed in different sessions along with the session chairs and audience.

The conference was held in hybrid mode. The plenary keynote lecture was presented in online mode and the link was provided for all the participants who are present in the conference and who are participated in online mode. A conference hall was dedicated for the participants in physical mode with a screen and projector. Another keynote lecture on nonlinear analysis was presented in physical mode and the lecture was transmitted to the online audience through online meeting.

IOP Conf. Series: Materials Science and Engineering 1273 (2023) 011001 doi:10.1088/1757-899X/1273/1/011001

Out of the selected papers of the conference, some technical papers are discussed in the conference venue and most of the papers are presented in online mode due to travel constraints and to save the time of travel. Parallel sessions were planned to provide enough time for the speakers to explain and for the discussion session after the presentation. Session chairs were connected online and offline to organize the parallel sessions.
The conference was concluded with vote of thanks in the afternoon session on 12th November 2022.

Various committee members of NCSGE – 2022 is given below.

Technical Advisory Committee
Dr. Sanjay Kumar Shukla, Associate Professor, Edith Cowan University, Australia.
Dr. Hemanta Hazarika, Professor, Kyushu University, Japan.
Dr. G.L. Sivakumar Babu, Professor, IISc., Bangalore.
Dr. Gali Madhavi Latha, Professor, IISc., Bangalore.
Dr. Nihar Ranjan Patra, Professor, IIT Kanpur.
Dr. S.K. Duggal, Professor, MNIT Allahabad.
Dr. A. Murali Krishna, Professor, IIT Tirupati.
Dr. Sreedeep S, Professor, IIT Guwahati.
Dr. Arvind Kumar Jha, Assistant Professor, IIT Patna.
Dr. Amit Kumar Das, Assistant Professor, NIT Silchar.
Dr. Pramod Kumar Gupta, Professor, IIT Roorkee.
Dr. KaustubhDasgupta, Associate Professor, IIT Guwahati.
Dr. Kumar Pallav, Associate Professor, CPUT, South Africa.
Dr. Vithal R. Tekumalla, Director, Floater Technology, IDARE, Houston, Texas.
Dr. N. Darga Kumar, Assistant Professor, JNTU, Hyderabad.
Dr. A. Cinitha, Senior Scientist, CSIR-SERC, Chennai.
Dr. G. Aruna Kumari, Civil Structural Engineer, Exxonmobil Services & Technology Pvt. Ltd., Bangalore.

Organizing Committee

Chief Patron:
Prof. Karunesh Kumar Shukla, Director, NIT Jamshedpur.
Patron:
Prof. R.V. Sharma, Deputy Director, NIT Jamshedpur.
Chairman:
Dr. Sanjay Kumar, HOD, Civil Engg., NIT Jamshedpur.
Convener:
Prof. Anil Kumar Choudhary and Dr. Virendra Kumar, NIT Jamshedpur.
Conference Organizing Secretary(s) / Chair(s):
Dr. Keshav Kumar Sharma, NIT Jamshedpur.
Dr. Awdhesh Kr. Choudhary, NIT Jamshedpur.
Dr. Seeram Madhuri, NIT Jamshedpur.
Internal Advisory Committee:
Dr.C.M.Rao, Dean R&C, NIT Jamshedpur.
Prof. M.K. Paswan, Dean Faculty, NIT Jamshedpur.

IOP Conf. Series: Materials Science and Engineering 1273 (2023) 011001 doi:10.1088/1757-899X/1273/1/011001

All faculty members, Dept. of Civil Engg., NIT Jamshedpur.

Publication Chairs:

Dr. A.K. Sinha and Dr. Sangeeta Kumari, NIT Jamshedpur.

IOP Conf. Series: Materials Science and Engineering 1273 (2023) 011002 doi:10.1088/1757-899X/1273/1/011002

Peer Review Statement

All papers published in this volume have been reviewed through processes administered by the Editors. Reviews were conducted by expert referees to the professional and scientific standards expected of a proceedings journal published by IOP Publishing Publishing.

- **Type of peer review:** Double Anonymous
- **Conference submission management system:** Morressier
- **Number of submissions received:** 35
- **Number of submissions sent for review:** 34
- **Number of submissions accepted:** 19
- **Acceptance Rate (Submissions Accepted / Submissions Received × 100):** 54.3
- **Average number of reviews per paper:** 2
- **Total number of reviewers involved:** 35
- **Contact person for queries:**
  Name: S. Madhuri
  Email: smadhuri.ce@nitjsr.ac.in
  Affiliation: Department of Civil Engineering, National Institute of Technology Jamshedpur

IOP Conf. Series: Materials Science and Engineering 1273 (2023) 012001 doi:10.1088/1757-899X/1273/1/012001

Rail Structure Interaction Analysis for Providing CWR on Un-Ballasted Multi-Span Open Web Steel Girder Bridge

D Jeetendra Prakash[1], P Sravana[2], and P Srinivasa Rao[3]

[1] Research scholar, Department of Civil Engineering, JNTU, Hyderabad, India
[2] Professor, Department of Civil Engineering, JNTUHCEH, Kukatpally, Hyderabad, India
[3] Professor, Department of Civil Engineering, JNTUHCEH, Kukatpally, Hyderabad, India
E-mail: dasarijittu@yahoo.co.in

Abstract. Rail structure interaction (RSI) requires thorough study to understand the force transfer mechanism between the track and bridge components due to train vertical live loads, braking/traction loads and thermal loads accordingly. RSI analysis gives us an outline of the axial stress in rail, support reactions transferred to the bridge components and the relative displacement caused between the track and bridge. This paper investigates rail structure interaction analysis of unballasted multi-span steel through girder railway bridge carrying dedicated freight corridor (DFC) of 32.5t axle load. For this purpose, RSI analysis is performed as per the guidelines given in UIC774-3R [7] Code of Practice (International Union of Railways) for two different multi-span bridge with a span configuration of 7x63m and 6x48m respectively and the axial stress in rail, relative displacement between rail and bridge and support reactions transferred to the bridge components are calculated based on the span and stiffnesses of the bridges. From the analysis results, it shall be concluded whether to explore the possibility of providing a provision of continuous welded rail (CWR) on the bridge or to adopt switch expansion joints (SEJ) or other expansion devices accordingly.

Keywords: Rail structure interaction (RSI), International union of railways (UIC), Dedicated freight corridor (DFC), Continuous weld rail (CWR), Long welded rail (LWR), Switch expansion joint (SEJ).

1. Introduction

The challenging tasks on design of railway bridges concerning to higher speeds & heavier loads are continuously increasing to fulfill the present and future demands. The safety of the track system under increased static and dynamic loads needs extra investigations which can be achieved with RSI analysis. RSI is influenced by many factors such as structural properties of the bridge, rail cross section, stiffness of super structure, stiffness of substructure, traffic speed, type of loading, type of support conditions etc., Detailed research on the phenomenon of RSI due to train live load and thermal loads on bridges has been ongoing almost since the use of bridges as a form of commercial transport. This research has been conducted across the countries, however there the focus is on several studies. Some studied the results of RSI analysis with higher train speeds and lesser axle loads on ballasted decks with different span configuration, but the present paper illustrates a unique aspect i.e., the study of RSI analysis on unballasted multi-span steel through girder bridges carrying very heavy axle load of 32.5t. For this purpose, initially the bridge is being analyzed for RSI considering DFC loading. Secondly, the axial stresses in the rail are examined to provide a provision of CWR on bridge. Finally, the lateral relative displacement of track and bridge are examined for train live loads acting horizontally and support reactions are evaluated at each abutment and pier location accordingly. Provision of CWR on bridge reduces the excessive vibrations which in-turn increases the passenger ride comfort, reduces derailment risks, reduces rail fractures at joints and essentially reduces very high maintenance cost accordingly. Therefore, in line with the results obtained from RSI analysis the provision of providing CWR or SEJ on bridge shall be concluded accordingly.

IOP Conf. Series: Materials Science and Engineering 1273 (2023) 012001 doi:10.1088/1757-899X/1273/1/012001

2. Scope and Objective

The scope and objective of this study are to illustrate the RSI analysis results of two different multi-span bridges and conclude the provision of providing CWR/SEJ on the bridge.

The following loading conditions are considered to perform RSI analysis of the proposed bridge as per the guidelines given in UIC774-3R [7] as shown in Figure.1

Load case 1: Traction/Braking loads
Load case 2: Temperature
Load case 3: Train vertical loads.

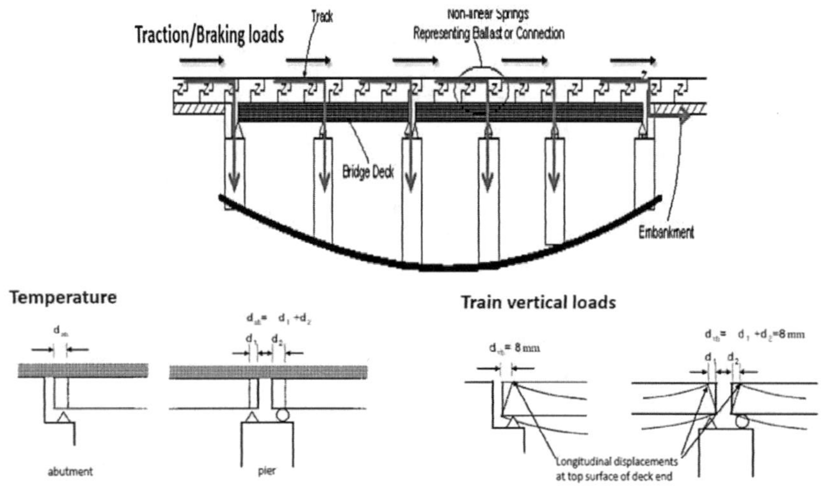

Figure 1. Illustration of structural system for calculating interaction effects using UIC774-3

Bridge parameters considered in the present research work are as given below.

- Stiffness of substructure
- Stiffness of superstructure
- Track stiffness on bridge
- Track stiffness on the embankment
- End rotation of the deck
- Vertical deflection of the girder
- Thermal loads
- Braking/Traction loads
- Train vertical live load

Finally, based on the above-considered parameters the axial stress in rail, the relative displacement between the rail and bridge components and support reactions are calculated as per the guidelines given in UIC774-3R [7] to propose a provision for providing CWR on the bridge or to adopt SEJ.

3. Design Criteria

Prior to embarking upon the numerical analysis of the CWR track-bridge system for the interaction effects, it is necessary to understand the important parameters which effect this phenomenon.

IOP Conf. Series: Materials Science and Engineering 1273 (2023) 012001 doi:10.1088/1757-899X/1273/1/012001

3.1. Support Stiffness

The entire support stiffness comprises of stiffness of each component of the bridge such as the Abutment, Pier, Piles and Soil in which they are embedded as shown in Figure.2

The support stiffness K including its foundation and the horizontal displacement parallel to the axis of the bridge is given as follows

$$K = \frac{H[KN]}{\Sigma \delta i[cm]} \tag{1}$$

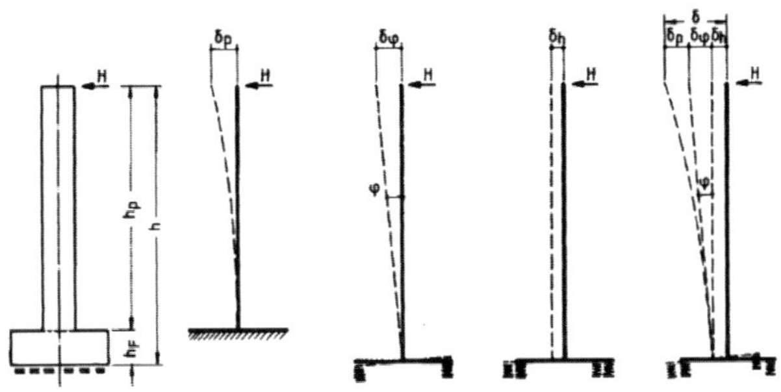

Figure 2. Factors influencing support stiffness (k)

With $\delta_i = \delta_p + \delta_\emptyset + \delta_h$

Where:

δ_p= Horizontal displacement at the head of the support due to elastic deformation.
$\delta_\emptyset$= Horizontal displacement at the head of the support due to rotation of the foundation
δ_h= Horizontal displacement at the support due to horizontal movement of the foundation.

3.2. End rotation of the Bridge

Train traffic loads cause the bridge deck to bend vertically, which in turn produces end rotation of the bridge deck as shown in Figure.3. The bending stiffness is a parameter of vertical deformation of the deck due to flexural displacement of the deck in the horizontal direction. Defining the correct cross-sectional area and the moment of inertia for the frame element defines the stiffness of the element in the corresponding directions.

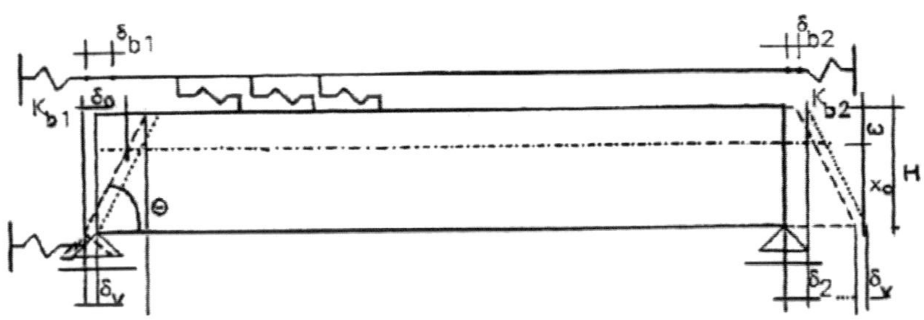

Figure 3. Effect of deck bending on the end sections

4. Bridge Geometry

4.1. Bridge details

Type of Bridge	= Railway Bridge	
Type of loading	= DFC Loading	
Axle load	= 32.5 t	
No. of Tracks	= Single Track	
Type of superstructure	= Open Web Steel Girder	
Type of sub-structure	= Abutments and Piers	
Type of foundation	= Piles	
Bridge-1 span configuration	= 7x 63.7	m
Bridge-2 span configuration	= 6x 48.0	m
Grade of concrete for the substructure, (fck)	= 35	N/mm^2
Grade of concrete for piles, (fck)	= 40	N/mm^2
Young's modulus of concrete for the substructure, $E_c = 5000\sqrt{fck}$	= 29580.4	N/mm^2
Young's modulus of concrete for piles, $E_c = 5000\sqrt{fck}$	= 31622.7	N/mm^2
Young's modulus of steel, (E_s)	= 200000	N/mm^2

4.2. Input data for track stiffness

Longitudinal Displacement of Rail considered [7]	= 0.5	mm
Resistance of Loaded Track [7]	= 60	kN/m
Resistance of Un-Loaded Track [7]	= 40	kN/m

Table 1. Open Web Steel Girder Bridge Details of Bridge-1 and Bridge-2

Description	Bridge-1 (7x63.7 m) (Mts)	Bridge-2 (6x48.0 m) (Mts)
Bridge span	63.7	48
No of spans	7 No's	6 No's
The total depth of OWG	9.84	9.703
Distance from C.G of OWG to the bottom chord	4.92	4.83
Distance from bottom chord to deck top	1.49	1.43
Distance from bottom chord to rail top (H)	1.79	1.63
Omega, (ω)	-3.43	-3.25
gamma, ($\Upsilon = \omega/H$)	-1.91	-2

Table 2. Abutment and Pier Details of Bridge-1 and Bridge-2

Description	Bridge-1 (7x 63.7 m) (Mts)	Bridge-2 (6x 48.0 m) (Mts)
Length of the abutment (Across traffic)	7.6	7.6
Width of the abutment (Along traffic)	1.35	1.1
Height of the abutment	7.18	6.71
Moment of inertia of Abutment	1.56 m^4	0.84 m^4
Length of the pier (Across traffic)	6	5.8
Length of the pier (Across traffic)	2	2.2
Width of the pier (Along traffic)	2	2.2
Height of the pier	13.23	6.28
Moment of inertia of Pier	4.79 m^4	6.3 m^4

Table 3. Pile Properties at Abutment Locations of Bridge-1 and Bridge-2

Description	Bridge-1 (7x 63.7 m)	Bridge-2 (6x 48.0 m)
	(Mts)	(Mts)
Diameter of the pile	1.5	1.5
Depth of fixity	8.27	9.4
Movable standing length	14.695	6.07
Total number of piles at the abutment	18 No's	12 No's

Table 4. Pile Properties at Pier Location of Bridge-1 and Bridge-2

Description	Bridge-1 (7x 63.7 m)	Bridge-2 (6x 48.0 m)
	(Mts)	(Mts)
Diameter of the pile	1.5	1.5
Depth of fixity	8.27	9.4
Movable standing length	14.695	6.07
Total number of piles at the abutment	18 No's	12 No's

Table 5. Stiffness of Abutment and Pier of Bridge-1 and Bridge-2

Description	Bridge-1 (7x 63.7 m)	Bridge-2 (6x 48.0 m)
	(kN/mm)	(kN/mm)
Stiffness of abutment	138.33	275.10
Stiffness of pier	51.26	97.41

Table 6. Stiffness of Abutment and Pier in Terms of Span of Bridge-1 and Bridge-2

Description	Bridge-1 (7x 63.7 m)	Bridge-2 (6x 48.0 m)
Stiffness of abutment	2.2L	5.82L
Stiffness of pier	0.81L	2.02L

Table 7. End Rotation of Girder for Bridge-1 and Bridge-2

Description	Bridge-1 (7x 63.7 m)	Bridge-2 (6x 48.0 m)
Train live load intensity	120.5 kN/m	122.7 kN/m
End rotation 'Θ' of the deck = $\frac{wL^3}{24EI}$	0.00184 rad	0.0017 rad
δ(ΘH) for deck	2.7 mm	1.521 mm

Table 8. Loads considered for RSI Analysis on Bridge-1 and Bridge-2

Description	Bridge-1 (7x 63.7 m)	Bridge-2 (6x 48.0 m)
Braking load intensity	18.7 kN/m	26.16 kN/m
Live load intensity of train acting vertically downward	120.5 kN/m	122.7 kN/m
Variation of temperature	35^0c	35^0c

IOP Conf. Series: Materials Science and Engineering 1273 (2023) 012001 doi:10.1088/1757-899X/1273/1/012001

5. Axial Stresses in Rail Calculated as Per UIC774-3R [7]

Note: Axial stresses in rail are calculated using graphs given in Appendix A and Appendix B of UIC774-3R [7]

Table 9. Axial Stresses in rail at Fixed Support with the Provision of CWR for Bridge-1 and Bridge-2

Loading condition	Description	Bridge-1 (7x 63.7 m)	Bridge-2 (6x 48.0 m)
		Axial stress N/mm^2	Axial stress N/mm^2
Braking load	Axial stresses in Rail at fixed support for "K60"	30.29	24.3
Temperature load	Axial stresses in Rail at fixed support for "K40" ΔT=35°	3.47	9.13
Train vertical load	Axial stresses in Rail at fixed support for "K60"	9.77	19.05

Table 10. Axial Stresses in rail calculated as per UIC774-3R [7] at Movable Support with the Provision of CWR for Bridge-1 and Bridge-2

Loading condition	Description	Bridge-1 (7x 63.7 m)	Bridge-2 (6x 48.0 m)
		Axial stress N/mm^2	Axial stress N/mm^2
Braking load	Axial stresses in Rail at fixed support for "K60"	30.29	25.45
Temperature load	Axial stresses in Rail at fixed support for "K40" ΔT=35°c	50.47	40.27
Train vertical load	Axial stresses in Rail at fixed support for "K60"	58.80	59.06

Table 11. Axial Stresses in rail calculated as per UIC774-3R [7] at fixed support with the Provision of SEJ for Bridge-1 and Bridge-2

Loading condition	Description	Bridge-1 (7x 63.7 m)	Bridge-2 (6x 48.0 m)
		Axial stress N/mm^2	Axial stress N/mm^2
Braking load	Axial stresses in Rail affixed support for "K60"	48.23	55.14
Temperature load	Axial stresses in Rail at fixed support for "K40" ΔT=35°c	0	0
Train vertical load	Axial stresses in Rail at fixed support for "K60"	0	0

IOP Conf. Series: Materials Science and Engineering 1273 (2023) 012001 doi:10.1088/1757-899X/1273/1/012001

6. Support Reactions Abutment and Pier are calculated as per UIC 774-3R [7]

The support reaction on Abutment and pier is calculated as per clause **3.3.3** given in UIC 774-3R for Bridge-1 and Bridge-2 and the results are tabulated in Table 12 and Table 13 respectively.

Table 12. Bridge-1: Summary of support reactions on Abutment and Pier for the load cases from RSI analysis

Location	Braking/Traction with CWR	Temperature Load	Train Vertical load	Braking/Traction with SEJ
	(kN)	(kN)	(kN)	(kN)
	(1)	(2)	(3)	(4)
Abutment	400	740	1174	1236
Pier- P1	315	148	0	1236
Pier- P2 to P5	315	74	0	1236
Pier- P6	315	370	274	1236

Table 13. Bridge-2: Summary of support reactions on Abutment and Pier for the load cases from RSI analysis

Location	Braking/Traction with CWR	Temperature Load	Train Vertical load	Braking/Traction with SEJ
	(kN)	(kN)	(kN)	(kN)
	(1)	(2)	(3)	(4)
Abutment	555.0	714.0	1097.0	1236.0
Pier- P1	451.3	142.8	0.0	1236.0
Pier- P2 to P3	451.3	0.0	0.0	1236.0
(n-1)th Pier- P4	451.3	71.4	0.0	1236.0
(n)th Pier- P5	451.3	357.0	548.5	1236.0

7. Verification of Horizontal displacement in abutment and pier for Braking loads with the provision of CWR and SEJ for Bridge-1 (7x 63.7m)

Note: Support reactions are calculated from the graphs given in Appendix A and Appendix B of UIC774-3R [7]

7.1. Horizontal displacement of Abutment with the provision of CWR

Maximum horizontal load due to braking in case of CWR	= 400	kN
Stiffness of the abutment	= 138.33	kN/mm
Horizontal displacement for braking load	= 3	mm
Allowable horizontal displacement [7]	= 5	mm

IOP Conf. Series: Materials Science and Engineering 1273 (2023) 012001 doi:10.1088/1757-899X/1273/1/012001

7.2. Horizontal displacement of Pier with the provision of CWR

Maximum horizontal load due to braking in case of CWR	= 315	kN
Stiffness of the pier	= 51.26	kN/mm
Verification of horizontal displacement for braking	= 6.145	mm
Allowable horizontal displacement [7]	= 5	mm

7.3. Horizontal displacement of Abutment with the provision of SEJ

Maximum horizontal load due to braking in case of SEJ	= 1236.0	kN
Stiffness of the abutment	=138.33	kN/mm
Horizontal displacement for braking load	= 9	mm
Allowable horizontal displacement [7]	= 30	mm

7.4. Horizontal displacement of Pier with the provision of SEJ

Maximum horizontal load due to braking in case of SEJ	= 1236.0	kN
Stiffness of the pier	= 51.26	kN/mm
Verification of horizontal displacement for braking	= 24.11	mm
Allowable horizontal displacement [7]	= 30	mm

8. Verification of Horizontal displacement in abutment and pier for Braking loads with the provision of CWR and SEJ for Bridge-2 (6x 48m)

Note: Support reactions are calculated from the graphs given in Appendix A and Appendix B of UIC774-3R [7]

8.1. Horizontal displacement of Abutment with the provision of CWR

Maximum horizontal load due to braking in case of CWR	= 555.0	kN
Stiffness of the abutment	= 275.10	kN/mm
Verification of horizontal displacement for braking	= 2.017	mm
Allowable horizontal displacement [7]	= 5	mm

8.2. Horizontal displacement of Pier with the provision of CWR

Maximum horizontal load due to braking in case of CWR	= 451.3	kN
Stiffness of the pier	= 97.41	kN/mm
Verification of horizontal displacement for braking	= 4.6	mm
Allowable horizontal displacement [7]	= 5	mm

8.3. Horizontal displacement of Abutment with the provision of SEJ

Maximum horizontal load due to braking in case of SEJ	= 1236.0	kN
Stiffness of the abutment	= 275.10	kN/mm
Verification of horizontal displacement for braking	= 4.5	mm
Allowable horizontal displacement [7]	= 30	mm

8.4. Horizontal displacement of Pier with the provision of SEJ

maximum horizontal load due to braking in case of SEJ	= 1236.0	kN
Stiffness of the pier	= 97.41	kN/mm
Verification of horizontal displacement for braking	= 12.7	mm
Allowable horizontal displacement [7]	= 30	mm

9. Results and Discussion

The present paper illustrates the results of the axial stress in rail, displacements of abutment/pier and support reactions at abutment/pier locations. For this study, an unballasted multi-span open web steel girder bridge having two different bridge span configurations of 7x63.0m and 6x48m respectively carrying DFC loading (32.5t Axle load) are analyzed using the guidelines given in UIC774-3R [7]. As per the guidelines given in UIC774-3R [7] the loads considered for the analysis are braking/traction loads, thermal loads and train vertical live loads as shown in Figure. 1. Based on the analysis results obtained, the provision of CWR or SEJ on the bridge is being adopted accordingly.
The axial stress in rail at the fixed support for braking/traction loads, thermal loads, and train vertical live loads with a provision of CWR for bridge-1 (7x63m) are 30.29 N/mm^2, 3.47 N/mm^2 and 9.77 N/mm^2 respectively and for bridge-2 (6x48m) are 24.3 N/mm^2, 9.13 N/mm^2 and 19.05 N/mm^2 respectively as shown in Figure. 4

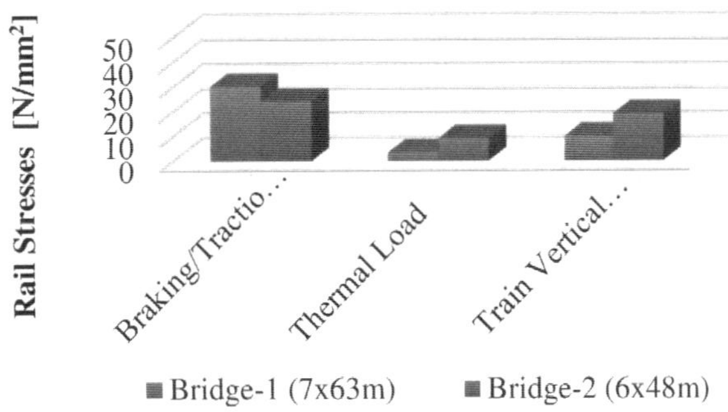

Figure 4. Axial Stress in Rail with Provision of CWR at Fixed Support for Bridge-1 and Bridge-2

The axial stress in rail at the fixed support for braking/traction, temperature, and vertical bending with a provision of CWR for bridge-1 (7x63m) are 30.29 N/mm^2, 50.47 N/mm^2 and 58.80 N/mm^2 respectively and for bridge-2 (6x48m) are 25.45 N/mm^2, 40.27 N/mm^2 and 59.06 N/mm^2 respectively as shown in Figure. 5

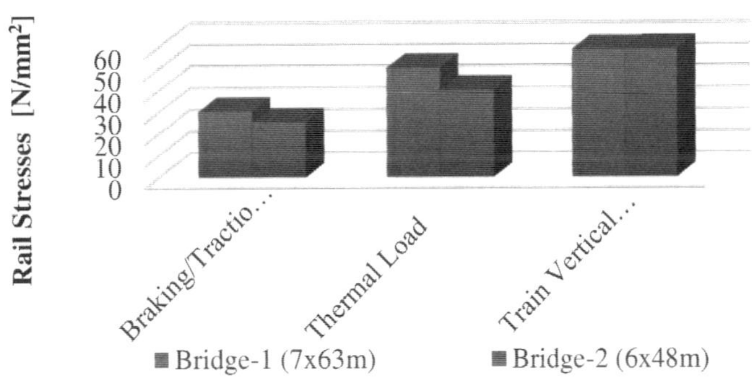

Figure 5. Axial Stress in Rail with Provision of CWR at Movable Support for Bridge-1 and Bridge-2

The axial stress in rail at the fixed support for braking/traction, temperature, and vertical bending with a provision of SEJ for bridge-1 (7x63m) are 48.23 N/mm^2, 0 N/mm^2 and 0 N/mm^2 respectively and for bridge-2 (6x48m) are 55.14 N/mm^2, 0 N/mm^2 and 0 N/mm^2 respectively as shown in Figure. 6 and Table-11.

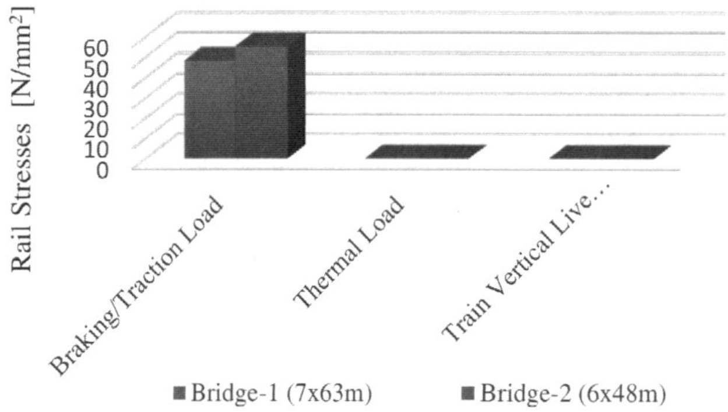

Figure 6. Axial Stress in Rail with Provision of SEJ at Fixed Support for Bridge-1 and Bridge-2

It may be noted that the total axial stresses in the rail for bridge-1 and bridge-2 with a provision of CWR at fixed support are 43.53 N/mm^2 and 52.48 N/mm^2 respectively as shown in Table-9 which are well within the allowable limit of 72 N/mm^2 as recommended in UIC 774-3R [7].
The total axial stresses in the rail for bridge-1 and bridge-2 with a provision of CWR at movable support are 139.56 N/mm^2 and 124.78 N/mm^2 respectively as shown in Table-10 which exceeds the allowable limit of 72 N/mm^2 as recommended in UIC 774-3R [7].
Hence a provision of SEJ at each support location shall be adopted which reduces the axial stress in the rail as shown in Fig. 6.
The summation of forces on the abutment and pier due to braking/traction loads, thermal loads, and train vertical Live load from the RSI analysis are tabulated in Table-12 for bridge-1 and Table-13 for bridge-2 respectively.
The horizontal displacement of the abutment and pier for bridge-1 considering braking/traction loads with a provision of CWR is 3 mm and 6.145 mm respectively as shown in Fig. 7(a).
The horizontal displacement of the abutment and pier for bridge-1 considering braking/traction loads with a provision of SEJ are 9 mm and 24.11 mm respectively as shown in Fig. 7(b).

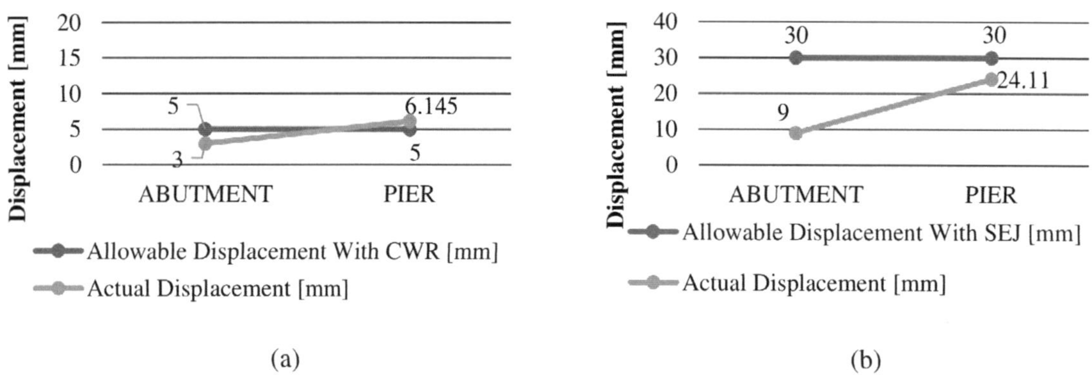

Figure 7. (a) Horizontal Displacement of Abutment and Pier with the provision of CWR for Bridge-1 (b) Horizontal Displacement of Abutment and Pier with the provision of SEJ for Bridge-1

The horizontal displacement of the abutment and pier for bridge-2 considering braking/traction loads with a provision of CWR is 2.017mm and 4.6mm respectively as shown in Figure. 8(a)
The horizontal displacement of the abutment and pier for bridge-2 considering braking/traction loads with a provision of SEJ is 4.5mm and 12.7mm respectively as shown in Figure. 8(b).

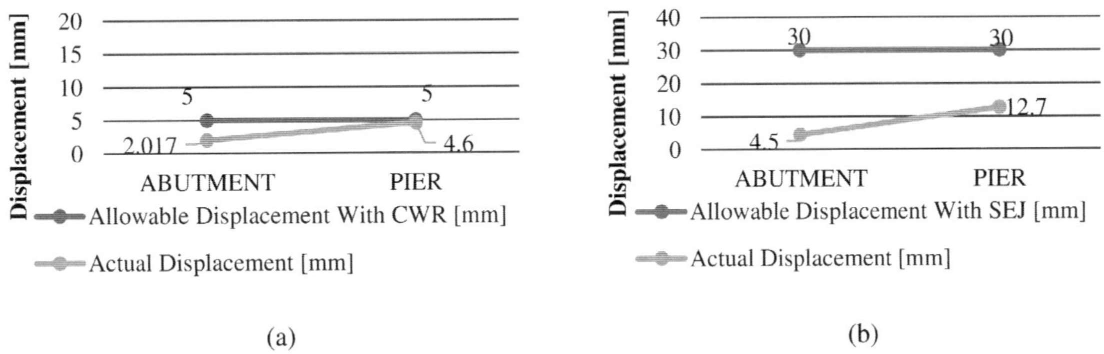

Figure 8. (a) Horizontal Displacement of Abutment and Pier with the provision of CWR for Bridge-2 (b) Horizontal Displacement of Abutment and Pier with the provision of SEJ for Bridge-2

It is observed that the horizontal displacement of the abutment for bridge-1 with a provision of CWR is within the allowable limit, however, the horizontal displacement of the pier exceeds the maximum allowable limit of 5mm.
It is observed that the horizontal displacement of the abutment and pier for bridge-1 with a provision of SEJ are within the allowable limits of 30mm.
It is observed that the horizontal displacement of the abutment and pier for bridge-2 with a provision of CWR and SEJ are within the allowable limit of 5 mm and 30mm respectively as specified in UIC774-3R [7].

10. Conclusion

- In Railway bridges, axial stresses in the rails, support reactions and vertical/horizontal displacements of the bridge components are envisaged due to interaction between the track and bridge, which are caused due to braking/ traction loads, thermal expansion/contraction of the bridge system, train vertical live loads. These interaction effects influence the design of the substructure and the super-structure of a bridge accordingly.
- It may be noted that axial compressive stresses in the rail increase drastically at the movable support with a provision of CWR, thus it is necessary to carry out an RSI analysis to check whether the axial stresses in the rail are within the allowable stress limits while providing a provision of CWR. However, the rail stresses can be reduced by means of providing SEJ with other expansion devices at required locations accordingly.
- Furthermore, it may be noted from the above study that the provision of CWR or SEJ on the bridge is not only influenced by the limits of axial stress in the rails but also depends on the longitudinal displacement limits of the bridge components accordingly.
- It may be observed from the above study that, the horizontal displacements due to braking/tractions loads are exceeding the allowable limits by providing CWR, however, the horizontal displacements are well within the allowable limits only by providing a provision of SEJ or other expansion devices accordingly.

IOP Conf. Series: Materials Science and Engineering 1273 (2023) 012001 doi:10.1088/1757-899X/1273/1/012001

References

[1] IRS Bridge Rules: 2008, Rules Specifying the Loads for the Design of Super Structure and Sub Structure of Bridges and For Assessment of the Strength of the Existing Bridge. (https://www.iricen.gov.in/iricen/Bridge_Manuals/BridgeRule.pdf)
[2] IRS Sub Structures & Foundation Code: 2013, Code of Practice for the Design of Sub Structures and Foundations of Bridges. (https://iricen.gov.in/iricen/Bridge_Manuals/FoundationSubstructureCode.pdf)
[3] IRS Concrete Bridge Code: 1997, Code of Practice for Plain, Reinforced & prestressed concrete for General Bridge Construction. (https://www.iricen.gov.in/iricen/Bridge_Manuals/CBCODE.pdf)
[4] IRS Steel Bridge Code: 2019, Code of practice for the Design of Steel or Wrought Iron Bridges Carrying Rail, Road, or Pedestal Traffic. (https://www.iricen.gov.in/iricen/Bridge_Manuals/Steel_Bridge_Code.pdf)
[5] IRC: 78-2014, Foundation and Sub Structure-Standard Specifications and Code of Practice for Road Bridges. (https://law.resource.org/pub/in/bis/irc/irc.gov.in.078.2014.pdf)
[6] RDSO Guidelines on Seismic Design of Railway Bridges Version 1, Nov 2015.
[7] UIC 774-3R-Rail Structure Interaction 2nd Edition, 2001.
[8] UIC 776-3R-Deformation of Bridges, 1st Edition, 1989.
[9] RDSO Guidelines for Carrying out Rail Structure Interaction Studies on Metro Systems Version 1, Nov 2014. (https://rdso.indianrailways.gov.in/works/uploads/File/RSI%20Guidelines.pdf).
[10] LWR on Ballasted Deck Bridges-UIC Approach.
[11] Shah, B. J., & Surti, S. K. (2015). The Need for Rail Structure Interaction Analysis for Bridges: A State-of-the-Art Review. IUP Journal of Structural Engineering, 8(3).
[12] Ahamed Ali Asif. K, Rajneesh K.Y and Sisir P. (2016). Strength Problems Associated with Track Bridge Interaction in Presence of Continuously Welded Rail. International Journal of Research Science & Management, ISSN: 2349-5197.
[13] McManus, A., Costa, L., & Gupta, A. (2017, April). Rail Structure Interaction for Multiple Span Railway Viaducts with Direct Fixed Track. In Austroads Bridge Conference, 10th, 2017, Melbourne, Victoria, Australia.
[14] Kumar, R., & Upadhyay, A. (2012). Effect of temperature gradient on track-bridge interaction. Interact Multiscale Mech, 5(1), 1-12.

IOP Conf. Series: Materials Science and Engineering 1273 (2023) 012002 doi:10.1088/1757-899X/1273/1/012002

Using Artificial Intelligence and Machine Learning Model for Prediction of Uniaxial Compressive Strength of GGBS Concrete

Bandaru Sri Sai Shanmukh[1] (0000-0002-4524-2113), Bandaru Sri Veera Bhadra Sasirekha[2] (0000-0002-3401-4465) and Varikuti Raja Sai Sampath[3]

[1]Department of Civil Engineering, University College of Engineering Kakinada (Autonomous), JNTUK, Kakinada, India
[2]Department of Electronics and Communications Engineering, Srinivasa Institute of Engineering and Technology, JNTUK, Kakinada, India
[3]Department of Advanced Construction Management, School of Construction, NICMAR, Hyderabad, India
E-mail: bshanmukh99@gmail.com

Abstract. Ground Granulated Blast-furnace Slag (GGBS) concrete is sustainable and also proven to show enhanced properties such as reduced heat of hydration, refinement of pore structure and increased resistance to corrosion and chemical attacks when compared with Ordinary Portland Cement concrete. Uniaxial compressive strength (UCS) and enhanced properties of the GGBS concrete are dependent on the proportions of GGBS and other elements in the mix. It is essential to accurately predict the UCS of the blended mix. Developments in computer hardware and the easy availability of research data made artificial intelligence and machine learning (AI & ML) prediction techniques feasible. In this study AI & ML techniques namely, linear regression, lasso regression and ridge regression are used to predict the UCS of GGBS concrete. Algorithms are trained using data collected from various standard publications. All data points are cleaned and then normalized with standard scalar function to avoid biased predictions. Root Mean Squared Error, Mean Squared Error, Mean Absolute Error and Coefficient of Determination for all three techniques are found to be almost identical. The above-mentioned AI & ML models have high prediction accuracy. Hence, based on the results AI & ML algorithms specified above can be reliably used in the above-mentioned GGBS concrete.

Keywords. Uniaxial Compressive Strength · GGBS Concrete · Artificial Intelligence and Machine Learning

1. Introduction

It is of no surprise to us that concrete is the most consumed commodity after water, in terms of per-capita consumption [1]. It is estimated that in 2020 alone the global demand of cement is 3000 million tons with a 4.7% growth per year. It is believed that 5-10% of the global anthropogenic CO_2 emissions is from cement manufacturers [2]. In the United States of America, of all the energy

IOP Conf. Series: Materials Science and Engineering 1273 (2023) 012002 doi:10.1088/1757-899X/1273/1/012002

produced nearly 0.6% was consumed in cement production [3]. Hence, it can be concluded that the production of cement in particular and concrete, in general, is carbon and energy intensive.

Governments throughout the world aiming to revive their economies from Covid-19 impacted economic slowdowns are largely focusing on infrastructure development [4]. Therefore, using conventional concrete in infrastructure development is not sustainable, as it emits large amounts of Greenhouse gases into the atmosphere every passing year with no sign of a decrease in the near future. Making concrete sustainable is the challenge. Many researchers and engineers have tried to make concrete sustainable. Using mineral admixture as cement replacement in concrete is one such novel technique.

GGBS is one such mineral admixture, which can be used as a cement replacement like fly ash (FA). This is largely produced by products from steel manufacturing industries. It is estimated that 7.8 million tons of blast furnace slag are produced in India [5]. At this point, it is to be noted that proper handling, management, and treatment of materials (GGBS, FA etc.) is poorly done and in most cases, disposal is done by dumping them in open places or in fills as heaps [6].

Research on GGBS blended concrete showed that it is sustainable concrete. GGBS blended concrete is also proven to show enhanced properties like reduced heat of hydration, resistance to corrosion and chemical attacks etc. when compared with Ordinary Portland Cement (OPC) concrete [5].

In the case of GGBS concrete UCS is dependent on proportions of GGBS and other elements in the mix accompanied by curing technique and age. To use GGBS concrete confidently in construction, it should have targeted UCS. Therefore, it is essential to accurately predict the UCS of the blended mix. Various statistical and accelerated testing techniques were developed over time to predict UCS, which are neither simple nor easy to use. In the meantime, there were huge developments in field computer hardware. The easy availability of research data and new computer hardware made AI & ML prediction techniques feasible.

2. Literature Review

Replacement levels of cement in concrete with 10% and 30% of GGBS are found to improve the average strength efficiency at 1.29 and 1.02 respectively at 28 days [7]. The further addition of GGBS beyond 55% of cement does not improve the UCS of GGBS concrete due to the presence of unreacted GGBS in the paste [8].

Initially, the pore volume of GGBS concrete is greater than that of ordinary concrete; however, with the age of concrete, the pore volume of GGBS concrete will gradually reduce. An increase in the GGBS ratio reduces the pore volume [9].

Under sustained loading corrosion resistance of reinforcement steel is found to be dependent primarily on crack width. However, the rapid chloride penetration test and permeability test showed that reinforcement steel in concrete with GGBS as a replacement to ordinary Portland cement has better corrosion resistance than ordinary concrete when exposed to 3.5% NaCl solution [10]. The UCS of GGBS concrete affected by hydrochloric acid (HCl) is found to be greater than the GGBS concrete affected to sulfuric Acid (H2SO4) [1]

Results of tests conducted on high-strength concrete with GGBS, broadly showed similar effects (workability, heat of hydration and strength) to those in lower-strength GGBS concrete, except in some cases [11]. Beneficial effects like higher UCS and sulfates resistance were noticed when GGBS is blended with other admixtures like FA to produce High-Durability Concrete [12]. For high-Performance concrete, to achieve given workability, GGBS significantly reduced the dosage of superplasticizer. [13]

To predict concrete UCS, in present times statistical regression methods are used [14]. With the help of empirical based models, it is difficult to formulate a precise equation [15]. To recognize relationships between UCS of GGBS concrete and input variables, AI&ML techniques are used in this study.

IOP Conf. Series: Materials Science and Engineering 1273 (2023) 012002 doi:10.1088/1757-899X/1273/1/012002

With advancements in the domain of computer technology and with more volumes of data being accessible, lead to the development of advanced techniques such as AI&ML. Research showed that practical applications of AI&ML happen to have large scope extended beyond theoretical and empirical studies of engineering problems [16].

Artificial neural networks (ANN) mimic the natural brain neural network [17]. Differing from traditional techniques of USC prediction, ANN does not employ any specific equation as input. In this technique, the ANN system can learn and train itself from given test data. It can continuously improve the predictions by supplying new sets of data. To improve prediction accuracy, the parameter condensation technique and weighting technique of input neurons were used [18]. Using ANN with the back propagation-learning algorithm for training, the results thus obtained when compared with actual laboratory test results proved that ANN is predicting the UCS accurately [19].

A study to predict the UCS of Waste Foundry Sand concrete found that ANN is less error-porn than multiple linear regression (MLR). With validation and uncertainty analyses, it is also noted that all models were within the permissible limits [20]. Fuzzy logic (FL) and ANN prediction values of High early-strength concrete UCS were compared in a study and concluded that ANN performed better [21].

Gene expression programming was considered in a study to predict the UCS of GGBS concrete. Mathematical equations, which are simple and practical, were proposed using the GEP algorithm to predict the UCS of GGBS concrete [22].

Table 1. Dataset statistics

Element	Min. value	Max. value	Mean	SD
Water (kg/m^3)	209	295	237.93	20.58
Cement (kg/m^3)	175	400	239.98	52.75
GGBS (kg/m^3)	0	440	148.28	130.9
FA (kg/m^3)	477	768	660.03	660.03
CA (kg/m^3)	209	295	1001.4	108.93
SP (kg/m^3)	0	0	0	0
Age (days)	7	365	110.085	119.04
UCS (MPa)	9.20	67.20	34.19	13.05

3. Data Set Description

Data is the most important component of this research. Data is given to the computer as the input, and algorithms process the data and give output. The present study has the input data with the following heads which are numerical representing proportional quantity of materials used for the manufacture of GGBS blended concrete: cement, water, fine aggregate, coarse aggregate, GGBS. Plasticizer or superplasticizer can be used to improve workability, which is also an input data. Further, number of curing days and USC are also given as input data. This data containing 224 samples [8] are collected from various standard publications. Table 1 has the statistics of the data used for this study. In Table 1 Mean and SD represents Arithmetic mean and Standard deviation respectively. To find correlation between two variables Pearson correlation coefficient is employed.

$$\rho_{xy} = \frac{\text{cov}(x, y)}{\sigma x \delta y} \tag{1}$$

In Eqs (1), cov (X, Y) is the covariance between X and Y. σX and σY are the standard deviation of X and Y respectively. Fig. 1 is a correlation matrix visualizing the correlation coefficients between variables. Correlation between two variables is seen in each cell of the correlation matrix.

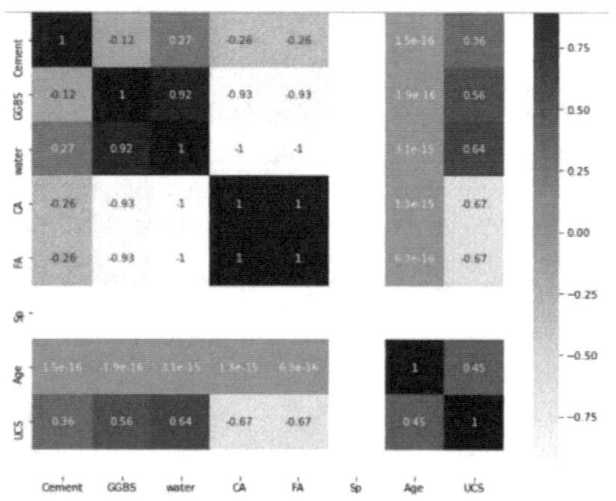

Figure 1. Correlation matrix.

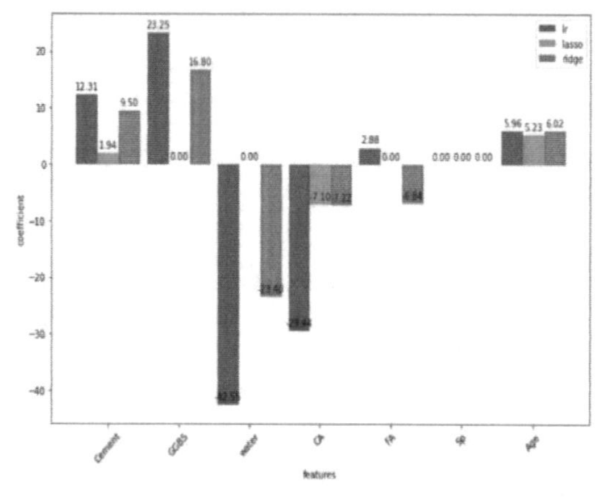

Figure 2. Features and coefficients of all three-regression techniques.

4. Methodology

4.1 Data Preparation

Data set having 224 data points was created referring various standard publications. All the data is first cleaned (removing the missing data and inaccurate data points in the considered data set) and then normalized with standard scalar function to avoid biased predictions. As the data collected has different units for different heads or tags (like kg/m^3 for cement and no. of days for age), there should be normalization process. In normalization process, various heads or tags are given a normalized coefficient to their units, which will give better results after the data processing stage without biased results.

4.2 Data Exploration

The prepared data, which is free from errors, is then used for data exploration. In this data exploration phase various statistical parameters are found for the data set of each data type and can be seen in Table 1. These parameters give an understanding of the data set. Various graphs such as pair plot and correlation matrix (heat map) of variables can be developed. These pair plot and heat maps make visualization of relationship between the different variables easily.

4.3 Data Processing and Predictions

Actual AI & ML algorithms are used in this stage for processing the data and giving the predictions. Data set is divided as a sample set and test set. Training of the algorithms is done by sample data set and testing set is used for results validation purpose. Evaluation of the results are done using performance measures and prediction are obtained. In this study linear regression, lasso regression and ridge regression are used to predict the UCS of GGBS blended concrete.

Linear regression is a technique used for modelling the relation between dependent and independent variables. Simple linear regression has one explanatory variable [23]. Linear estimate is defined by

$$y_i = \beta_0 + \sum_{j=1}^{p} x_j \beta_j \qquad (2)$$

Lasso regression is a technique where data values are shrunk towards a central point. Models with fewer parameters are encouraged in lasso regression models. Hence, for simple cases lasso regression is the best estimator [23]. Lasso estimate is defined by

$$\beta^{lasso} = min_{\beta_0,\beta} \left\{ \sum_{i=1}^{N} \left(y_i - \beta_0 - x_i^T \beta_j \right)^2 \right\} \text{ subjected to } \sum_{j=1}^{p} |\beta_j| \leq t \quad (3)$$

Rigde regression is used to estimate the coefficients of multiple-regression models, when linearly independent variables are highly correlated. Its variance and mean square estimator are mostly smaller than the least square estimators [23]. Ridge estimate is defined by

$$\beta^{ridge} = min_{\beta} \left\{ \sum_{i=1}^{N} \left(y_i - \beta_0 - x_i^T \beta_j \right)^2 \right\} \text{ subjected to } \sum_{j=1}^{p} \beta_j^2 \leq t \quad (4)$$

Let the sample consists of N cases, each of which consists of p covariates and a single outcome. Let y_i be the outcome and $x_i = (x_1, x_2 \ldots, x_p)_i^T$ be the covariate vector for the ith case. Here β_0 is the constant coefficient $\beta = (\beta_1, \beta_2, \ldots, \beta_p)$ is the coefficient vector, and t is a pre-specified free parameter that determines the degree of regularization.

4.4 Performance Evaluation Methods

For the evaluation of the predictive performance of models, the following performance measures are considered

RMSE is the standard deviation of the prediction errors (residuals). It shows how data is concentrated around the best-fit line. [15, 24]

$$\text{RMSE} = \sqrt{\frac{1}{m} \sum_{i=1}^{m} (P_i - T_i)^2} \quad (5)$$

MSE measures the average of the squares of the errors [21]. It is also called as mean squared deviation (MSD)

$$\text{MSE} = \frac{1}{m} \sum_{i=1}^{m} (T_i - P_i)^2 \quad (6)$$

MAE measures the errors between observation pairs. In the current context comparisons of predicted versus observed. [15, 24]

$$\text{MAE} = \frac{1}{m} \sum_{i=1}^{m} |P_i - T_i| \quad (7)$$

R-squared (R2) measures the proportion of the variation in the dependent variable that is predictable from the independent variable(s). [24] It is also known as Coefficient of Determination

$$R^2 = 1 - \frac{\sum_{i=1}^{m} (P_i - T_i)^2}{\sum_{i=1}^{m} (T_i - T^*)^2} \quad (8)$$

Where P_i is the Predicted values and T_i is the Tested values, mean value of all the tested values is given by T^* and the total number of the data points is m.

Table 2. Performance evaluation measures.

Data	Training Data				Test Data			
Model	RMSE	MSE	MAE	R2	RMSE	MSE	MAE	R2
Linear regression	5.54	30.74	4.52	0.82	6.72	45.1	5.75	0.67
Lasso regression	7.4	54.77	5.94	0.69	7.37	54.27	6.05	0.61
Ridge regression	5.88	54.52	4.72	0.8	6.52	42.46	5.5	0.69

5. Results And Discussion

As stated earlier, to predict the UCS of GGBS concrete, AI & ML techniques are used extensively in this study Eqs. (2), (3) and (4). Figure 3 and 4 shows predicted versus true strengths of GGBS concrete of all three AI & ML techniques considered in this study. Performance evaluation measures such as RMSE, MSE, MAE and R2 can be seen in Table 2 are calculated using Eqs (5), (6), (7) and (8).

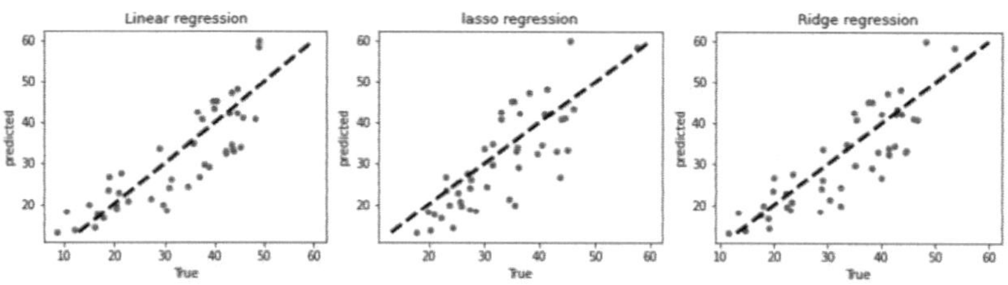

Figure 3. Predicted versus true strengths for training data

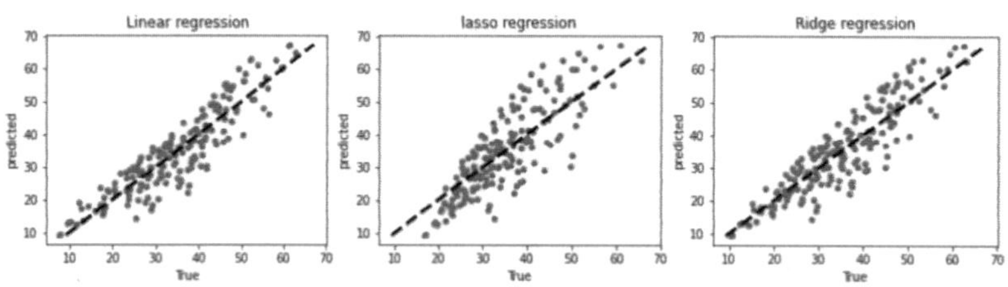

Figure 4. Predicted versus true strengths for test data

A comparison of Figures 3 and 4 shows that for all three AI & ML techniques considered in this study had almost similar RMSE, MSE, MAE and R2 values, indicating UCS prediction can be carried out by the regression techniques used in this study.

From figure 2 it is to be noted that despite having similar performance evaluation values, lasso regression has lower coefficients meaning it reduces the complexity of the model. It can also be observed that coefficients of linear and ridge regression are almost the same.

Figures 3 and 4 shows the predicted and true UCS values by all three AI & ML techniques considered. Wherein performance of all AI & ML techniques considered in this study seem to be similar. Hence, AI & ML techniques considered in this study can be used for determining the UCS of GGBS concrete.

6. Conclusions

In this study to predict UCS of GGBS concrete linear regression, lasso regression and ridge regression techniques are used.

- All three regression techniques considered in this study can accurately predict the UCS of GGBS concrete. The performance all three regression techniques considered are comparable.

- The complexity of the model is reduced in Lasso regression, as it has lower coefficients for given features.
- The three regression techniques considered in this study have good extrapolation capabilities in predicting the UCS of GGBS concrete.

Acknowledgements
We express our deep gratitude to our Guruji Leela Venkata Satya Siva Sai for his support in developing computer code. We thank, Prof. Dr Lakshmi Veerni of the Department of Civil Engineering, University College of Engineering (Autonomous), Kakinada, JNTUK for encouraging us. We are deeply obliged to our friends Samanchi V.S.L. Manisri Gayathri, Mahanti Keerthana Naidu and Basava Pavan Putra Reddy for their constant help and support throughout the project work.

References
[1] Karri, S. K., Rao, G. R., & Raju, P. M. (2015). Strength and durability studies on GGBS concrete. SSRG International Journal of Civil Engineering (SSRG-IJCE), 2(10), 34-41
[2] Ighalo, J. O., & Adeniyi, A. G. (2020). A perspective on environmental sustainability in the cement industry. Waste Disposal & Sustainable Energy, 2(3), 161-164.
[3] Rodrigues, F. A., & Joekes, I. (2011). Cement industry: sustainability, challenges and perspectives. Environmental Chemistry Letters, 9(2), 151-166
[4] Abhijit Mukhopadhyay (2021). Post-pandemic economic recovery: seven priorities for India. ORF Occasional Paper No. 295, Observer Research Foundation.
[5] Shetty, M. S., & Jain, A. K. (2019). Concrete Technology (Theory and Practice), 8e. S. Chand Publishing.
[6] Bandaru Sri Sai Shanmukh & Bagadi Saikumar (2020). A study on the efficacy of RHA as a full replacement for clay in brick-blocks. International Research Journal of Engineering and Technology (IRJET), 7(3), 4927-4933.
[7] Babu, K. G., & Kumar, V. S. R. (2000). Efficiency of GGBS in concrete. Cement and Concrete Research, 30(7), 1031-1036.
[8] Oner, A. D. N. A. N., & Akyuz, S. (2007). An experimental study on optimum usage of GGBS for the compressive strength of concrete. Cement and concrete composites, 29(6), 505-514.
[9] Cheng, A. S., Yen, T., Liu, Y. W., & Sheen, Y. N. (2008). Relation between porosity and compressive strength of slag concrete. In Structures Congress 2008: Crossing Borders (pp. 1-8).
[10] Cheng, A., Huang, R., Wu, J. K., & Chen, C. H. (2005). Influence of GGBS on durability and corrosion behavior of reinforced concrete. Materials Chemistry and Physics, 93(2-3), 404-411.
[11] Domone, P. L., & Soutsos, M. N. (1995). Properties of high-strength concrete mixes containing PFA and ggbs. Magazine of Concrete Research, 47(173), 355-367.
[12] Liu, S., Zhu, M., Ding, X., Ren, Z., Zhao, S., Zhao, M., & Dang, J. (2021). High-durability concrete with supplementary cementitious admixtures used in corrosive environments. Crystals, 11(2), 196.
[13] Djellouli, H., Aitcin, P. C., & Chaalaar, O. (1990). Use of Ground Granulated Slag in High-Performance Concrete. Special Publication, 121, 351-368.
[14] Bhanja, S., & Sengupta, B. (2002). Investigations on the compressive strength of silica fume concrete using statistical methods. Cement and Concrete Research, 32(9), 1391-1394.
[15] Chou, J. S., & Pham, A. D. (2013). Enhanced artificial intelligence for ensemble approach to predicting high performance concrete compressive strength. Construction and Building Materials, 49, 554-563.

[16] Reich, Y. (1997). Machine learning techniques for civil engineering problems. Computer-Aided Civil and Infrastructure Engineering, 12(4), 295-310.

[17] Hoła, J., & Schabowicz, K. (2005). New technique of nondestructive assessment of concrete strength using artificial intelligence. NDT & E International, 38(4), 251-259.

[18] Lee, S. C. (2003). Prediction of concrete strength using artificial neural networks. Engineering structures, 25(7), 849-857.

[19] Gupta, R., Kewalramani, M. A., & Goel, A. (2006). Prediction of concrete strength using neural-expert system. Journal of materials in civil engineering, 18(3), 462-466.

[20] Tavana Amlashi, A., Alidoust, P., Pazhouhi, M., Pourrostami Niavol, K., Khabiri, S., & Ghanizadeh, A. R. (2021). AI-based formulation for mechanical and workability properties of eco-friendly concrete made by waste foundry sand. Journal of Materials in Civil Engineering, 33(4), 04021038

[21] Reddy, P. N., & Naqash, J. A. (2019). Strength prediction of high early strength concrete by artificial intelligence. Int J Eng Adv Technol, 8(3), 330-334.

[22] Shahmansouri, A. A., Bengar, H. A., & Ghanbari, S. (2020). Compressive strength prediction of eco-efficient GGBS-based geopolymer concrete using GEP method. Journal of Building Engineering, 31, 101326.

[23] Hastie, T., Tibshirani, R., Friedman, J. H., & Friedman, J. H. (2009). The elements of statistical learning: data mining, inference, and prediction (Vol. 2, pp. 1-758). New York: springer.

[24] Feng, D. C., Liu, Z. T., Wang, X. D., Chen, Y., Chang, J. Q., Wei, D. F., & Jiang, Z. M. (2020). Machine learning-based compressive strength prediction for concrete: An adaptive boosting approach. Construction and Building Materials, 230, 117000.

Thermal Analysis of Reinforced Beam-Column Joint

Utkarsh Agnihotri [1](0000-0003-4593-0431), Vinayak Mishra [2](0000-0002-4742-6549), and Singh R. P. [3] (0000-0001-5409-9024)

[1]M.Tech. Structural Eng., Civil Engineering Department, Institute of Engineering and Technology, Lucknow
[2]Assistant Professor, Civil Engineering Department, Institute of Engineering and Technology, Lucknow
[3]Chief Engineer, Central Public Works Department, New Delhi, India
agnihotriutsav1998@gmail.com

Abstract. In concrete structures, temperature changes can cause strains of the same order of magnitude as live and dead loads in some circumstances. When thermal expansion or contraction is controlled, thermal stresses are induced. This report discusses the performance of reinforced joints under thermal loading conditions. Here we analysed the variance in stress and deflection when the temperature increases. The joints are modelled in ETABS and imported into ANSYS for further thermal analysis. The aim is to evaluate the RC beam-column joints at high temperatures in transient response. The results discuss how RC joints respond to temperature variations. The deformations with temperature variation are examined at high-temperature conditions.

Keywords: ETABS Software · RC Beam-Column Joint · Thermal Analysis · ANSYS Software · ISO 834 Fire Curve · Finite Element Analysis.

1 Introduction

Concrete is the most widely used material in any civil engineering construction, from building to bridges. The benefits of using concrete are its cost-effectiveness, speed of construction, availability, etc. But one of the often-ignored benefits of using concrete is the protection it offers from fire. It offers enhanced protection from the fire hazard in the structure.

The behaviour of concrete structures subjected to heat stress is a critical practical concern. Indeed, there is a substantial quantity of research on the behaviour of plain concrete as well as RCC at increased temperatures[1].

High-temperature exposure is a severe situation that causes changes in material characteristics, which might influence the overall behaviour of a structural element[2]. The behaviour of each element is extremely important in the study of fire hazards in the building. Fire hazards in a building section can have serious consequences since its failure might collapse the entire structure [3][4].

Normal fire affects RC members by increasing the temperature of the concrete mass. Temperature rise lowers the mechanical characteristics of steel and concrete. They may also cause explosive spalling of concrete member surface fragments. Concrete, on the other hand, can suffer considerable damage when subjected to extended fire exposure or very high temperatures [5].

The primary goal of this research was to look at the effect on beam column joint during fire circumstances. A finite-element analysis was done to examine the thermal behaviour of the RC beam column joint.

To account for the previous studies conducted in the same domain, here are a few noteworthy mentions: As per previous study by **N. Parthasarathi et.al. (2020)** the major goal under transient state as well as study state at higher temperature is evaluated in RC beam column junction. The variation in a temperature is carried out under various loading gradually ranging from 40%- 60% of the collapse load. Estimation of the variation of temperature inside the tested sample is recorded through a thermocouple of K type at regular interval. Deformation in the rotation at junction as temperature rises and failure patterns are examined at higher temperature[4].

Essam H. El-Tayeb et.al. (2015) Here, behaviour demonstrated by RC beam elements and framed structures under thermal stress, including dead and service loads, is investigated in this research to analyse the effects of change in temperature. Temperature gradients of various types, including uniform, linear, and nonlinear, are addressed. Analysis is done in ABAQUS through finite element approach.

The research study findings show the modelling of concrete structure has significant effect on these structures under temperature variation. The behaviour of beams and frames varies greatly depending on change in the gradient of temperature under nonlinear is studied through realistic profile [3].

An attempt has been made by **Xudong Shi et.al. (2004)** Six specimens with varying concrete cover thicknesses were evaluated to see how the clear cover affected the characteristics of reinforced concrete flexural elements subjected to fire. These test findings reveal that the bottom clear cover has a substantial impact on the prototype final loading capability, although the intensity of the influence shown as above decreases in relation with the CC thickness increases. As a result, unnecessarily increasing the bottom clear cover to improve model fire resistance is not recommended. When compared to the bottom clear cover, the lateral CC has a less influence on the specimen fire resistance. This work proposes the idea of similar CC thickness to better estimate the influence of clear cover on the flexural strength of RC elements subjected to fire [6].

By Dietmar Hosser et.al. (1994) The advancement of adequate computer modelling systems to anticipate the reaction of structures to fire hence now allows individual structures to be analysed in light of actual fire-exposure circumstances. Experiments on simply supported composite beams attached to reinforced concrete slabs had been conducted at the University of Braunschweig. The good agreement between experimental and computation findings shows the effectiveness of finite-element theory in estimating the fire-resistant properties of beams. The impact of the top transverse reinforcement used in the concrete slabs for the fire resistance and fracture propagation is investigated. The use of the findings of building fire safety design is also examined [7].

R. Srinivasan and K. Sathiya et.al. (2010) The Structural components, moulded in concrete come in a variety of shapes and sizes in structures. Understanding how the components in question respond during the loading process is critical to developing an overall proper and efficient structure. Experimental testing has been widely applied to analysis individual parts as well as the impacts of concrete strength under loads. While this process provides real-life responses, it is very time consuming and tedious considering the materials used can be rather expensive. Finite element analysis has also been utilised to examine these components.

The research based investigation seeks to evaluate the findings of an elastic analysis of the reinforced concrete beam having transverse loading done with an analysis software to those obtained from a standard theoretical analysis [8].

2 Research significance

The behaviour of the RC structures in temperature variation is studied at its peak is illustrated according to the designing code. Under fire hazard beam having effective cover are according to the IS code. Under cyclic loading the study of the RC structure is tedious job thus deformation at junction under high thermal stresses can be easily analyzed finite element method. The various cyclic loading such as dead, live and fire loadings with certain combinations can be analysis.

IOP Conf. Series: Materials Science and Engineering 1273 (2023) 012003 doi:10.1088/1757-899X/1273/1/012003

Figure 1(a). Beam column joint

Figure 1(b). Reinforcement Detail

3 Objective

To model RC beam column joint in ETABS & study the effect of fire hazard on beam column joint and the structure through temperature variation using ANSYS.

4 Methodology

The study will be done in two parts. First DESIGNING and second one is ANALYSIS. In designing stage, we will design residential building with the help of software (ETABS). After designing thermal analysis of a joint is done with the help of ANSYS. The design data of the ETABS is used to model the beam column joint in ANSYS then ISO 834 fire curve is applied to the model an analyse the results.

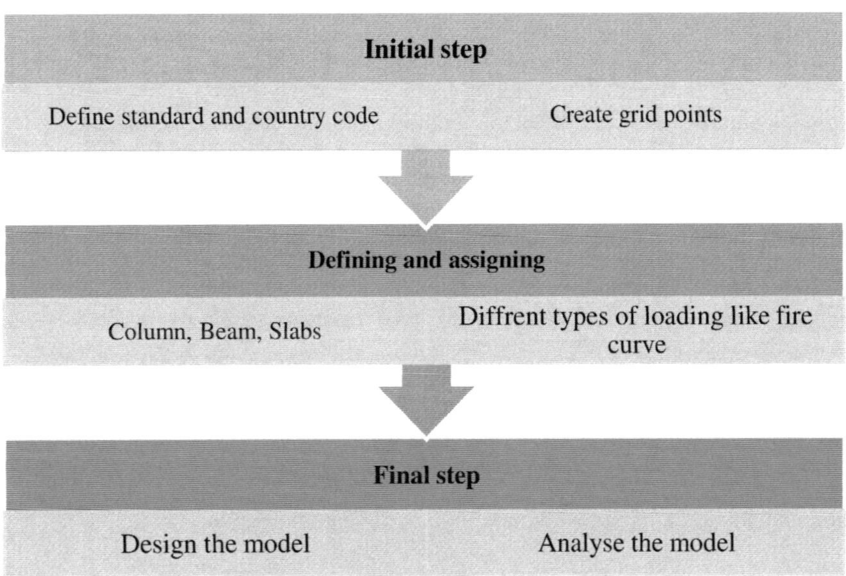

Figure 2. Methodology flow chart

5 Building parameters

Table 1. Material Properties

Name	Type	Design strength	E(Mpa)
Concrete	M30	F_c= 30 Mpa	27386.13
HYSD	Fe 415	F_y = 415	200000

Table 2. Building Parameters

Area of the building	20.1*12.3 m^2	
Height of the building	18m	
Column	450*300 mm^2	
Beam	300*450 mm^2	
Slab type	Conventional slab	
Slab thickness	125mm	
Wall	Periphery	250 mm
	Internal	150mm
AAC block	Unit weight	6 kn/m^2
	Compressive strength	4 kn/m^2
Plaster	Unit weight	25 kn/m^2
	Thickness	25 mm

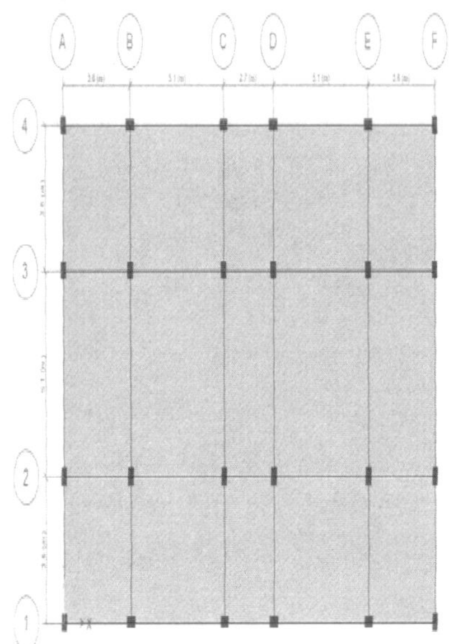

Figure 3. Plan view of model

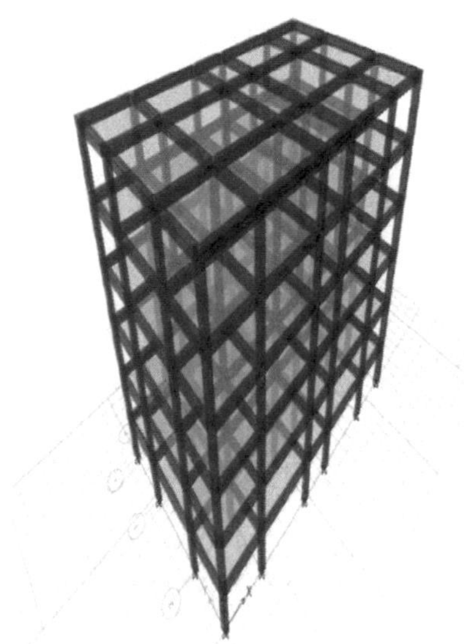

Figure 4. 3D view of building

6 Modeling

Previous researchers used ABAQUS software to create a finite element analysis model to evaluate the deflection behaviour, first and ultimate fractures of a reinforced concrete beam column junction exposed to point load loading.

Figure 5. Meshing

The same procedures are utilised in various software and at varying temperatures (transient state) across time. Here we perform our analysis in **ANSYS**. The beam length is 3600 mm, with a C/S of 300 * 450 mm^2, and the column height is 3000 mm, with a cross section of 450 *300 mm^2. The specimen is made of hexahedron mesh, and contacts between surfaces is employed between the conjunctions of steel with concrete. As illustrated in **Fig. 4**, the mesh pattern for the considered beam - to - column joint present in the analytical model.

7 Method of analysis

Finite element Software (ANSYS) is employed for the analysis. The model is investigated in order to obtain the structural behaviour of an RC beam and column joint at an elevated temperature in a real-world scenario. The model has been designed in ETABS and analyzed through ANSYS. The prototype model employs transient state analysis for simulation.

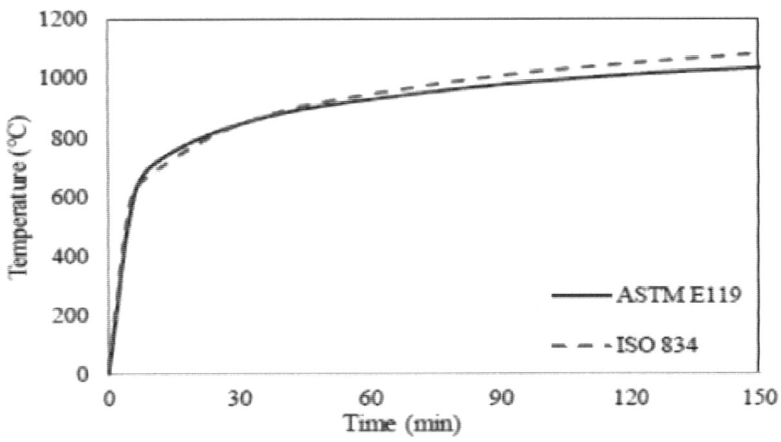

Figure 6. Fire curve (ISO 834)

8 Results and Discussion

As previously stated, fire breakout in a building lead to variable thermal gradient being applied across all elements of the building, which may lead to a rapid transit of heat flow through the building and as a result the heat eventually spreads evenly over the entire surface.

However, in practice, the heat penetration varies with respect to a plethora of other variables such as:

- Grade of concrete
- Location of reinforcement
- Material of reinforced
- Abundance of reinforcement
- Thickness of CC

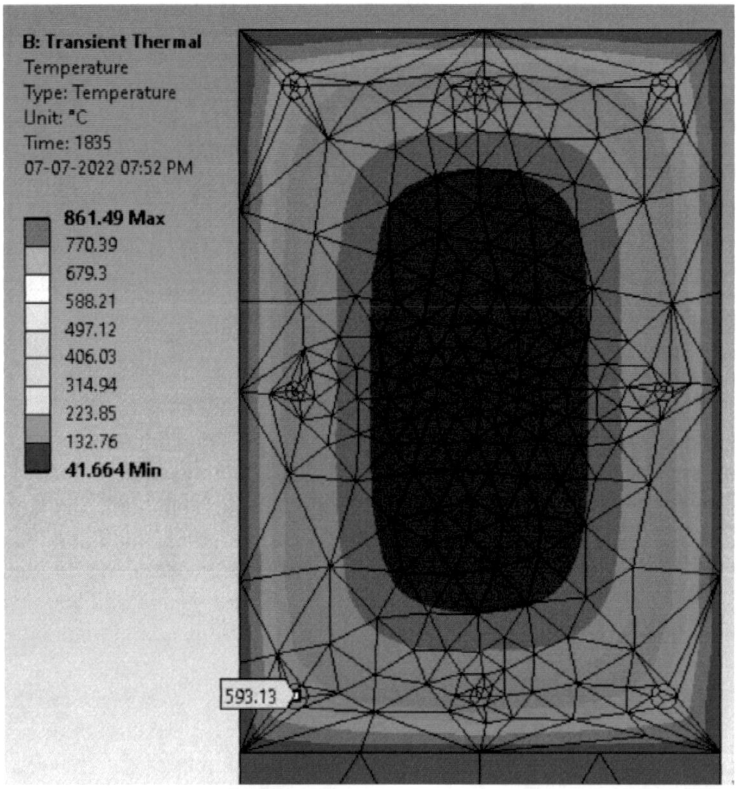

Figure 7. Column cross section

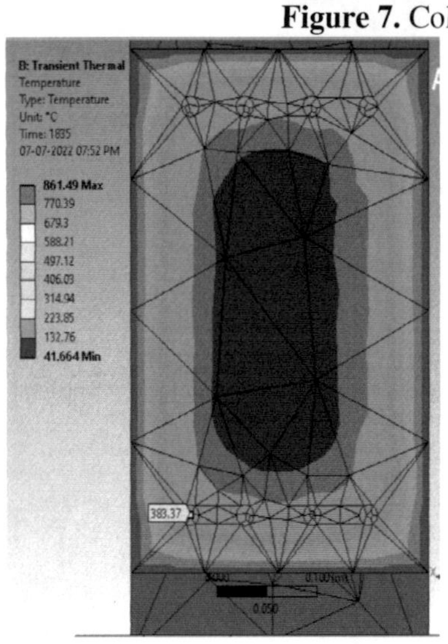

Figure 8. Beam cross section

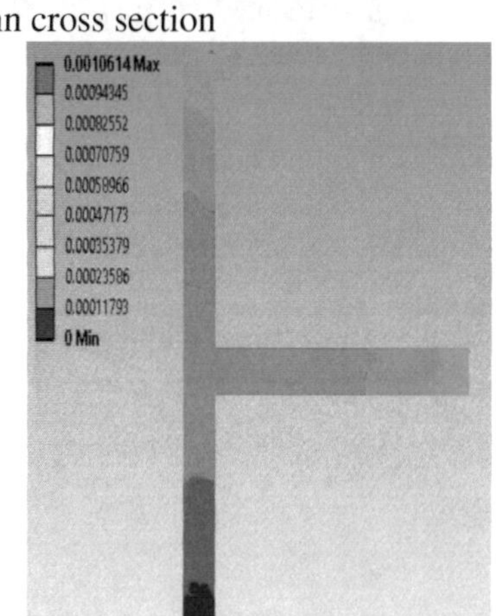

Figure 9. Deformation in structure

As stated above, the section is homogeneous, the temperature distribution varies w.r.t. the clear cover and reinforcement, other variables being kept constant.
It was hence observed in the simulation that in **1835 seconds,** the temperature of reinforcement in the column reached 593°C while that of the beam remained lower at 383°C.
Since the section is taken homogeneous, therefore the only responsible attribute for the above thermal variation is clear cover. The beam has larger value of clear cover leading to a slower rate of heat transfer.
It is to be duly noted that the temperature is applied in accordance to the ISO 834 fire curve.
Maximum deflection according to IS 456:2000 is = span/350 = 10.285mm.
Maximum deflection occurred due to thermal stress 1.0614mm.
It is therefore evident from the above numerical values that the beam-column interface is not affected by the applied heat stresses.

9 Conclusion

The RC beam column joint is studied and compared at high temperatures in transient states. The RC finite element model is thermally loaded using the ISO 834 fire curve.
If the beam column joint is exposed to fire from all sides, then the steel temperature reaches 593°C at **30 minutes 35 seconds.** Beyond this temperature failure of reinforcement is likely to occur.
The other point of the study is that the clear cover plays an important role to protect reinforcement from fire. In beam clear cover is 50mm and in column 40mm. So, at same time the temperature of reinforcement in the column and beam is 593.13 °C and 383.37°C respectively.
During a fire hazard thermal failure criterion is more significant than deflection criteria.
In our study we obtained that the maximum deflection of out model is in permissible limits i.e., 10.285mm so it shows that the applied heat stresses does not affect the RC beam column joint.

References

[1] R. De Borst and P. P. J. M. Peeters, "Analysis of concrete structures under thermal loading Rene de Borst and Paul P.J.M. Peeters," 1989.
[2] F. George, A. Balaji, and A. Raj, "Fire resistant analysis of t-beam and L-beam," vol. 3, no. 4, pp. 1148–1155, 2015.
[3] E. H. El-tayeb, S. E. El-metwally, H. S. Askar, and A. M. Yousef, "Thermal analysis of reinforced concrete beams and frames," Hous. Build. Natl. Res. Cent., 2015, doi: 10.1016/j.hbrcj.2015.02.001.
[4] Parthasarathi, N., Satyanarayanan, K. S., & Thamilarasu, V. (2021). Performance of reinforced concrete beam column joint under high temperature. Materials Today: Proceedings, 40, 52-55., doi: 10.1016/j.matpr.2020.03.497
[5] B. Jacob, A. Balaji, and E. John, "Open Access Behaviour of concrete structures under fire - a comparative study between IS 456 : 2000 and finite element software ANSYS American Journal of Engineering Research (AJER)," pp. 62–66, 2013.
[6] X. Shi, T.-H. Tan, K.-H. Tan, and Z. Guo, "Influence of Concrete Cover on Fire Resistance of Reinforced Concrete Flexural Members," J. Struct. Eng., vol. 130, no. 8, pp. 1225–1232, 2004, doi: 10.1061/(asce)0733-9445(2004)130:8(1225).
[7] Osama EI-Nesr Dietmar Hosser, 1 Thomas Dorn, "Experimental and numerical studies of composite beams exposed to fire By Dietmar Hosser, 1 Thomas Dorn, 2 and Osama EI-Nesr 3," vol. 120, no. 10, pp. 2871–2892, 1995.
[8] R. Srinivasan and K. Sathiya, "Flexural Behavior of Reinforced Concrete Beams Using Finite Element Analysis (Elastic Analysis)," Bul. Institutului Politeh. Din Iaşi, Construcţii. Ărhitectură, no. Lx, pp. 31–41, 2010.
[9] IS 456, "Plain Concrete and Reinforced," Bur. Indian Stand. Dehli, pp. 1–114, 2000.
[10] IS 3809, "Fire Resistance Test of Structures," Bur. Indian Stand., pp. 1–19, 1979.
[11] M. Kisan, S. Sangathan, and J. Nehru, "IS 1642: 1989 Fire safety of buildings (general): Details of construction-code of practice, Bureau of Indian Standard, New Delhi," pp. 1–16, 1989.

Study of a Flat Slab Building with a Shear Wall at Core and X-Bracing

Akshansh singh[1] (0000-0001-8581-9135), Vinayak mishra[2] (0000-0002-4742-6549), Singh R. P.[3] (0000-0001-5409-9024)

[1](M.Tech. Structural Eng.), Civil Engineering Department, Institute of Engineering and Technology, Lucknow

[2]Assistant Professor, Civil Engineering Department, Institute of Engineering and Technology, Lucknow

[3]Chief Engineer, Central Public Works Department, New Delhi, India

singhakshansh000@gmail.com

Abstract. Construction with flat slabs is quite dangerous when there is strong earthquake shaking. Two different forms of flat slab buildings—flat slab with shear wall at core and flat slab with x-bracing—are considered in the study. These flat slab constructions' structural rigidity is increased by this lateral force-resisting mechanism. It has been found that adding a shear wall considerably boosts the flat slab structure's ability to withstand lateral forces.

Keywords: Shear wall, Core, Pushover, X-bracing, ETABS Software, Seismic Response.

1. Introduction

One of nature's most destructive occurrences is seismic activity, and it frequently happens without warning. In the modern world, traditional RC Frame constructions are widely used for building. For architectural flexibility, space allocation, and flexibility of formwork with variation in construction speed, flat slab construction offers substantial advantages over conventional RC Frame building. [1]. Since tectonic and volcanic activity is the main causes of earthquakes, it is our duty to conform strictly to the earthquake-resistant design requirements of IS1893 and IS13920. Two different forms of flat slab buildings are considered in this research. Two models were utilised with their behaviour of flat slab buildings studied, and the work uses a nonlinear static pushover analysis technique. [2].

Klemencic et al. conducted multiple experiments on the connections between post-tensioned walls and slabs, and the results showed that the joints tend to demonstrate failure at an internal storey drift of 5%.
Weidlinger and Ettouney examined the tectonic lateral disruption of a flat-slab high-rise in New York City.
Vinod Goud (2015): - The dimensional stability of shear wall-flat slab interface is researched; major aims include resistance to several forces of action as well as the benefits of shear walls on the performance of the building structures under consideration of excitation under the influence of seismic forces [3].
Robertson et al research Shear loads and uneven pressures transmitted between slabs and columns resulted in brittle punching failure. When compared with a flat slab with a drop and a shear wall at various locations during an earthquake, uneven moments cause significant shear stresses present in the slab [4].

2 Pushover Analysis

The method used to determine the non-linear behaviour of concrete structures subjected to dynamic loadings, such as maximum values of displacement, shear, overturning moment, minimum drift, and maximum drift in the storey. That is a type of non-linear static and dynamic analysis in which the strength of the structure is assessed beyond its proportionality limit.[5].

RCC flat slab systems are a typical type of reinforced concrete construction, mainly in mid-sized companies, with regard to their effective span/depth ratios and the respective savings obtained by limiting storey heights. In the absence of beams, the pressure formerly absorbed by the removed columns cannot be transferred, these slabs are more prone to progressive collapse than slab-beam-column complexes. As a result, evaluating the resistance to the progressive collapse of RC flat slab structures is crucial. Therefore, it was advised that flat-slab construction only be utilised as the solution for vertical structural members in buildings braced by framework or shear walls responsible for the structure's transverse capacity in seismically prone sites. To withstand gravity loads, the major lateral load-resisting structural parts' lateral elastic deformation must be absorbed by the slab-column interfaces to prevent punching failure.[6].

3 Aim

Elucidate the result of respective displacement, shear at the base, and drift in the respective storey are parameters in code IS-1893:2016 for a varied number of flat slab building models, as well as to optimize the loading capacity and behaviour of model 1 i.e., flat slab building with respect to a lateral force resisting elements, and analysing the behaviour of flat slab buildings facing impending seismic deformation through the nonlinear approach.

Table1. Elements and design

S. No.	Parameter	Dimension
1	Type	Commercial
2	Slab Type	Flat slab
3	Dimensions	25*15
4	Elevation	G+9
5	Ground storey height	3 meters
6	Height b/w floors	3 meters
7	Total height	30 meters
8	Slab thickness	150 mm
9	Drop thickness	200mm
10	Drop width	2 meters
11	The thickness of the shear wall	200 mm
12	Column	400*400 mm

Table 2. Material Properties (I.S. 456: 2000)

S. No.	Parameter	Dimension
1.	Grade of slab	M20
2.	Grade of column	M25
3.	Reinforcement	Fe 415

Table 3. Seismic Elements (I.S code 1893: 2016)

S. No.	Parameter	Dimension
1.	Earthquake Zone	V
2.	Ratio of damping	5%
3.	Importance factor	1
4.	Soil type	Type II (medium soil)
5.	Response reduction factor	5

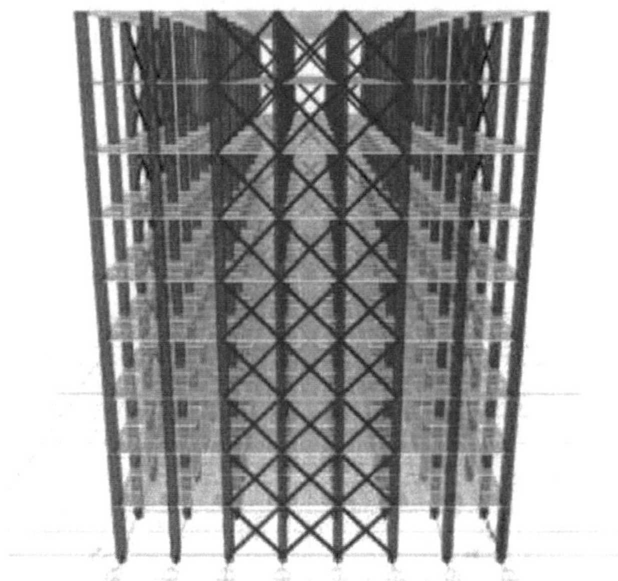

Figure 1. Model 1-flat slab with shear X-bracing

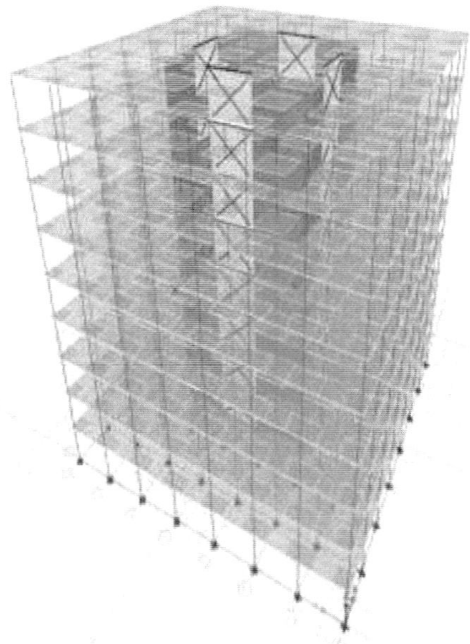

Figure 2. Model 1-flat slab wall at core

4. Methodology

With reference to this work, we design a G+9 RCC framework with a 25x15m plan layout. Following that, we apply material attributes, properties of the section, and condition of loading in the ETABSv16 programme under code IS 1893 (Part 1). The pushover analysis is then used to examine the behaviour which is non-linear and evaluate which is more economical & practically possible [7].

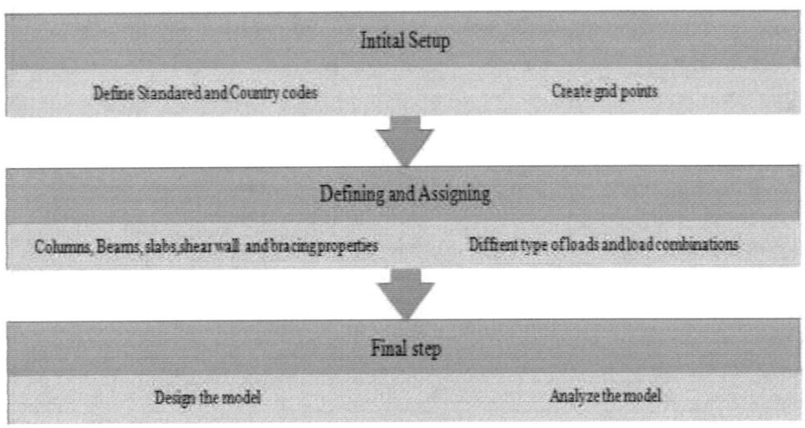

Figure 3. Methodology flow chart

5. Results and discussion

The comparison study was conducted under certain conditions

5.1 Overturning moment

The table shows the overturning moment within the sub and super-structure caused by rocking earthquake ground motion in both the horizontal and vertical planes. The overturning moment with a negative value represents the seismic forces causing a disturbance which needs to be counterbalanced, producing an increase in stability at the shear wall with x-bracing as compared to the shear wall at the core, and so model 1 is substantially more stable than model 2[8].

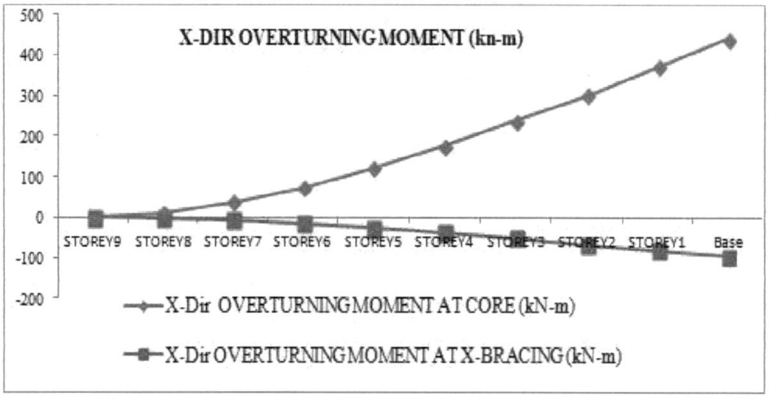

Figure 4. X –Direction overturning moment

Construction using X-Bracing has a lower overturning moment than building with a shear wall at the core. IS 456:2000, Cl. 20.1, states that the structure is stabilized under this case of a resisting moment which must be more than 1.2 the maximum flexure through the dead load, higher than 1.4 of loads applied [9].

5.2 Max storey Displacement

As illustrated in the table below, the shear walls at the centre and x-bracing are compared side by side with storey displacement as a parameter. The allowable value of storey displacement according to IS 1893:2016 is 0.4% of floor height (H). The shear wall, which acts as a restraining structure, restricts lateral displacement at extremely various locations. A comparison of the two is shown below.

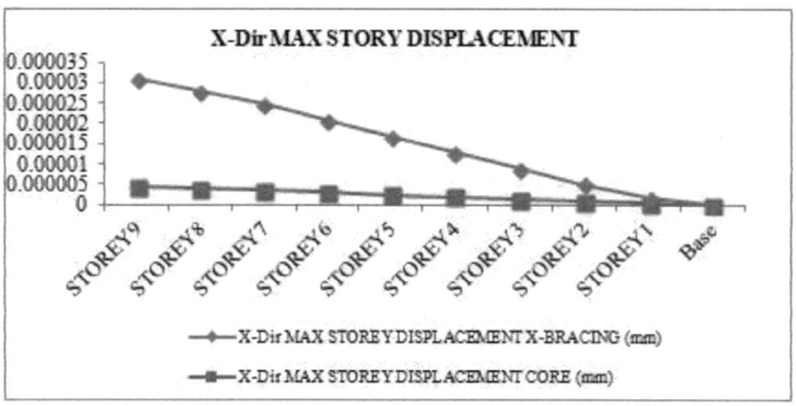

Figure 5. Max storey Displacement

The parameters such as displacement in the storey, and drift of the storey are smaller at the lowermost & more at the topmost floor, according to the charts below. According to IS 456:2000, Cl. 20.5, the average storey displacement of a flat slab should be under the permissible value of 0.2% of the height (H) for sway in the lateral direction at the topmost storey, here (H) total height of the structure. The minimum storey drift under any condition should not be larger than 0.4% times the storey height, according to IS code 1893 Part -I Cl. 7. 11.1[10].

The storey drift in a structure with shear walls at the core is 0.000341 m, but the storey drift in a building with X-bracing is 0.00085m. The maximum allowable limit is 0.192m. As a result, our structure is safe; drift values in both the y and x directions are within the allowable limit set by IS 1893[11].

5.3 Stiffness

The crucial parameters such as displacement feature must be managed to avoid damage to buildings exposed to ground accelerations brought on by earthquakes is the relative storey drift ratio.

The necessity for narrative drifts varies with ground motion's temporal variance. The relationship is inversely proportional to each other for stiffness and storey drift, the stiffer the structure, the smaller the storey drift in the relevant direction.

X-Bracing can be used to limit the flexibility of a concrete frame when the value of drift exceeds the allowable value of drift. The usage of shear walls increases the rigidity of the wall. The level drift decreases dramatically as the stiffness increases.

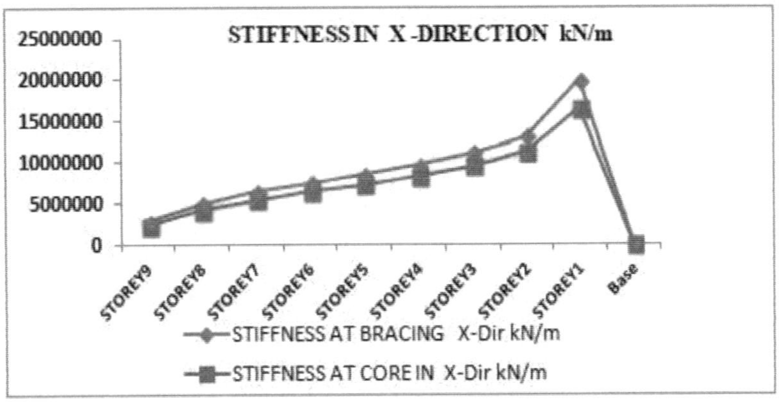

Figure 6. Stiffness

MODEL with a shear wall at the core has poorer stability when compared to MODEL with x-bracing. Thus, X-Bracing improves building stability under lateral stresses [12].

5.4 Storey shear

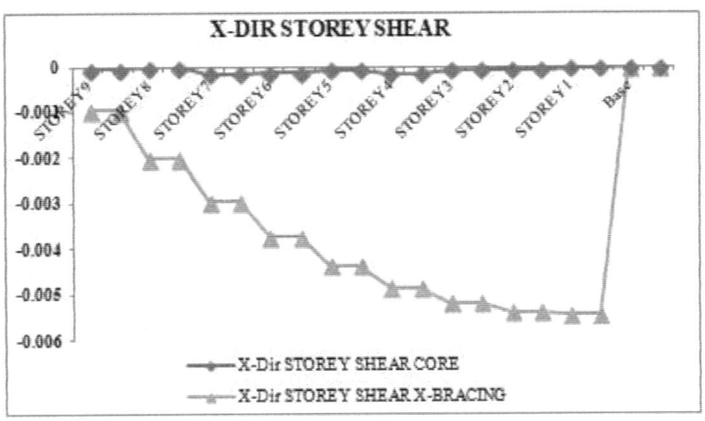

Fig. 7. Storey shear

The x direction is virtually nil for a shear wall at the core when storey shear is considered, although its maximum value for x bracing is 0.0054, which is likewise within the allowable limit.

6 Conclusion

Every storey is substantially smaller in the event of a shear wall at the core under the action of lateral force further the lateral force diminishes from top to bottom for X-bracing.

The shear wall at the core provides superior resistance to lateral pressures, boosting the structure's capacity to resist seismic disturbance under lateral forces and its stability.

Thus, the construction of a flat slab with an X-bracing resistance system provides a more flexible nature rather than shear wall construction at the centre or core. Earthquake lateral resisting systems in structural are solutions which can be taken as replacements, as per paper like shear walls and steel or X-bracing to reduce deformation which demands during big earthquakes. When compared to shear wall x-bracing, the value of displacement achieved under lateral force action for a shear wall at the core is economically superior.

References

[1] Pitroda, s. G., Nehru, j., & Sangathan, m. K. S. (2005). इंट रनेट मानक. Group, 1(2005).
[2] Tuken, a., & Siddiqui, n. A. (2015). Sbc-based assessment of shear wall quantity in moment resisting frame buildings. Ksce journal of civil engineering, 19(1), 188-199. Doi: 10.1007/s12205-014-0143-9.
[3] Joshi, d. D., & Murnal, p. B. (2013). Performance of flat slab structure using pushover analysis. J mech civ eng, 8(3). Doi: 10.9790/16840834144.
[4] Depeursinge, a., Racoceanu, d., Iavindrasana, j., Cohen, g., Platon, a., Poletti, p. A., & Müller, h. (2010). Fusing visual and clinical information for lung tissue classification in high-resolution computed tomography. Artificial intelligence in medicine, 50(1), 13-21.
[5] R. Klemencic, m. K. Associates, j. A. Fry, m. K. Associates, g. Hurtado, and C. Berkeley (2006). Performance of post-tensioned slab."
[6] Department of Defense (DoD). (2005). Unified facilities criteria (UFC), design of buildings to resist progressive collapse." unified Facility Criteria 4-023-03, dep. Defence, USA, no. July 2009, 2009.
[7] H. S. Kim and d. G. Lee, "efficient analysis of flat slab structures subjected to lateral loads," Eng. Struct. vol. 27, no. 2, pp. 251–263, 2005, doi: 10.1016/j.engstruct.2004.10.005.
[8] Junju Xie and Zenping wen, "a measure of drift demand for earthquake ground motions based on Timoshenko beam model," 14th world conf. Earthq. Eng., no. January, p. 13, 2008.
[9] K. Golghate, v. Baradiya, and a. Sharma, "pushover analysis of 4 storey' s reinforced concrete building," int. J. Latest trends eng. Technol., vol. 2, no. 3, pp. 80–84, 2013.
[10] D. Vamvatsikos and c. Allin cornell, "incremental dynamic analysis," Earthq. Eng. Struct. Dyn, vol. 31, no. 3, pp. 491–514, 2002, doi: 10.1002/eqe.141.
[11] S. Kokot, a. Anthoine, p. Negro, and g. Solomos, "static and dynamic analysis of a reinforced concrete flat slab frame building for progressive collapse," eng. Struct., vol. 40, pp. 205–217, 2012, doi: 10.1016/j.engstruct.2012.02.026.
[12] Rathod Chiranjeevi, Sabbineni Ramyakala, mandala Venugopal, and Nandanar Anusha, "seismic performance of flat slab with drop and conventional structure," int. J. Eng. Res., vol. V5, no. 10, pp. 283–295, 2016, doi: 10.17577/ijertv5is100240.

Seismic Site Response Analysis of MMMUT Gorakhpur of Uttar Pradesh, India

*Shubhamjeet Yadav[1], Madan Chandra Maurya[2], Keshav Kumar Sharma[3]

[1]Student, Master of Technology, Department of Civil engineering, Madan Mohan Malaviya University of Technology, Gorakhpur, India 273010
[2]Assistant Professor, Department of Civil engineering, Madan Mohan Malaviya University of Technology, Gorakhpur, India-273010
[3]Assistant Professor, Department of Civil Engineering, National Institute of Technology, Jamshedpur, India-831014
*Corresponding author. Email: mcmce@mmmut.ac.in, 2020013416@mmmut.ac.in

Abstract: An endeavour has been made to assess the spatial fluctuation of the profundity of endured and designing bedrock in Gorakhpur Uttar Pradesh, North India utilizing Multichannel Analysis of Surface Wave (MASW) study. One-layered MASW study has been done at Madan Mohan Malaviya University of Technology, Gorakhpur campus and Shear-wave Velocity $V_{S_{30}}$ are estimated. MASW overview at 3 location and a Standard Penetration Test SPT-N from the profound geotechnical boreholes data used for comparison of site classification. The deduction of this work might be utilized as contributions for seismic tremor risk the board by lessening the seriousness of earthquake shaking through plan of tremor Earthquake risk resilient structure strong designs. The data that collected from the MASW experimental setup is feed using the software ParkSEIS (v.3.0), we obtain Shear Wave Velocity (V_{S30}). The obtained $V_{S_{30}}$ is used to plot Response Spectra for MMMUT Gorakhpur. The Time History data used in this thesis is Nepal Earthquake 2015 data which is collected from Indian Metrological Department (IMD) Delhi, India. SPT- N value data is collected from Awas Vikas Parishad for the New Administration Building build at MMMUT campus Gorakhpur. The Time History data of Nepal Earthquake 2015 is analysed using DEEPSOIL v7 Software and a systematic layer wise study of MMMUT Campus soil is carried out, results such as Ground Motion Parameters, Response Spectra, Spectral acceleration vs Frequency plot, Tripartite Spectrum and Response spectra corresponding to 5% damping are obtained.

Keywords. Seismic Site Response, Site Classification, MASW, Shear wave velocity, MMMUT Gorakhpur.

1. Introduction

In several countries around the world, earthquakes are the most common natural disasters, which cause socioeconomic losses. The world has been encountering a tremendous number of risks through different natural disasters like floods, storms, Tsunamis, fires, and earthquakes. From a report from the international disasters database (EM-DAT), 57% casualty are due to earthquakes, and Asia and the pacific plate are badly influenced by earthquakes [2]. Literature shows that structural damage depends on the behaviour of the superstructure as well as the soil.

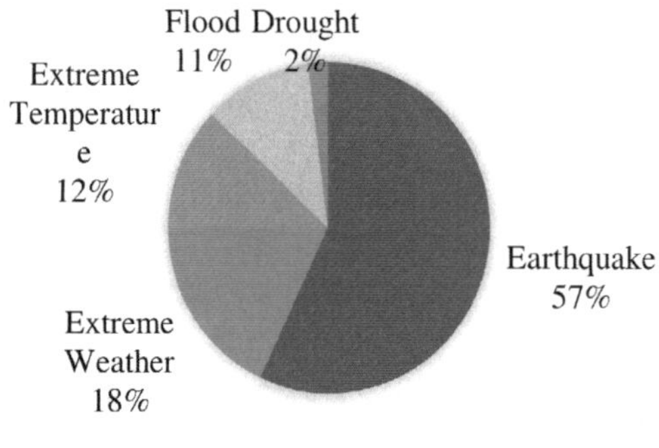

Figure 1. Casualty due to different natural disasters in last 20 year [2]

Another report from centre of research on the epidemiology of disasters (CRED) show that in last 20 years frequency of earthquake are increased as shown in fig. 2.

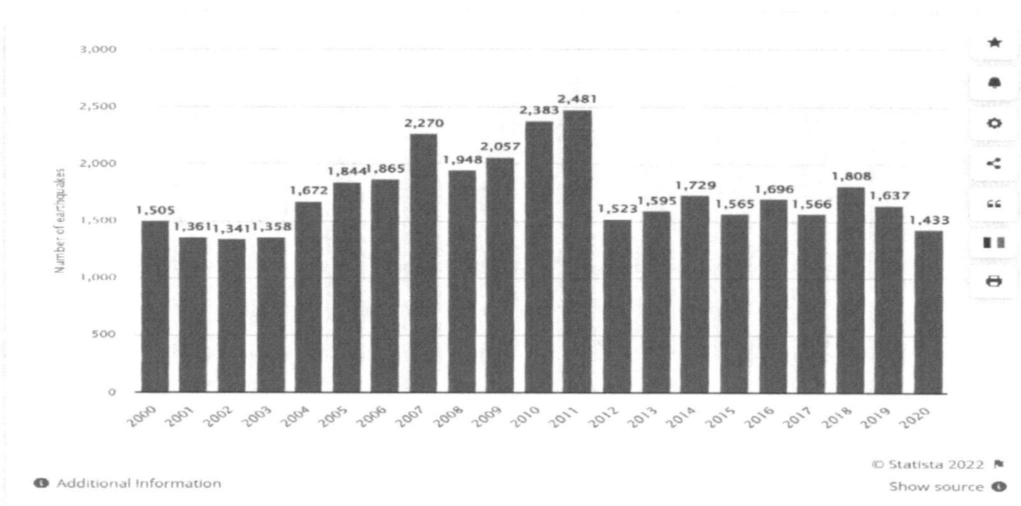

Figure 2. Number of earthquakes in last 20 years [3].

1.1 Damage During Past Earthquakes

Previous studies have shown that simply designing the superstructure to be seismically sound is insufficient to prevent the structure from collapsing during an earthquake [4]. Several damages have been documented during earthquakes as a result of a lack of understanding of the effect of soil structure interaction in the construction of structures and foundation system [5]. Whole structure can remain safe but due to failure of foundation it will overturn during earthquake Mexico earthquake, the highest buildings were demolished because of the soil foundation's breakdown of the bearing capacity [6]. The interaction effect was significant in the short boost of the natural period in the 1995 Kobe earthquake of intensity M=6.9 that resulted in the collapse [8].

IOP Conf. Series: Materials Science and Engineering 1273 (2023) 012005 doi:10.1088/1757-899X/1273/1/012005

The damage result causes by different Earthquake whether human loss or economy leads me to carry out this research work to find out the potential hazard of Earthquake in Gorakhpur region.

1.2 Classification of Seismic Site

Seismic codes have become notable throughout recent years due to no unusual place quakes worldwide. The site impacts address seismic ground reaction propensities and arrangements of seismic code. The craving for fitting, versatile design spectra are concurring with illustrations of soil [9]. The seismic force is the best method for representing net seismic site online impacts for designing errands and for wealthy purposes like micro zonation shown in this [10]. The present codes of seismic in Europe, America (2009, EC8; Global Building Code, UBC 97; NEHRP) delivered significant information furthermore, work incorporated the net seismic site on-line impacts dependent for the most part upon significant exploratory and hypothetical outcomes [11]. The precise soil classification is presented dependent for the most part upon a superior depiction of soil profiles utilizing popular geotechnical boundaries like pliancy file (PI), undrained shear strength (Su) what's more, not unusual place shear wave speed (Vs) values [12].

The major contribution of the proposed work:

- This research presents a simple approach that allows design engineers to address the effects of Response Spectra in the seismic design of normal mid-rise structures, ensuring design safety and dependability, also response of different layer of soil is part of this study.
- This study is used to figure out how the Response Spectra gives the vital contribution in designing of multi-story buildings in terms of the factors like Seismic Waves generated during earthquake to better understand the Seismology of Gorakhpur City responds to dynamic loading and to investigate and develop an interaction scenario formulating the interfaces between soil and building foundation contacts in order to better understand the behavior of the soil system during a seismic event.

This work is structured as follows: The related work in the field of seismic site response analysis is presented in the section II. Furthermore, section III provides the methodology proposed work. Section IV gives the experimental analysis of the proposed work and section V gives the results obtained while performing the experiments of the proposed work. Lastly section VI depicts the conclusion of the proposed work.

2. Literature Background

MASW is a seismic approach which can be used for geotechnical characterization of near-surface materials [13]. The MASW has been located to be an extra environment friendly approach for unravelling the shallow subsurface properties [14]. In the field of geotechnical engineering, MASW is widely used in earthquake for site response and micro zonation studies [15]. In particular, MASW is utilized to measure shear-wave pace and dynamic properties and identify spatial variants of velocity of shear-wave and boundaries of subsurface material [16].

Site response assessments employing geotechnical and geophysical experiments and relationships between N_{60} and G_{max} in accordance to specific soil properties of the site worked on attempt to compare ground response of soil using geotechnical and geophysical methods in **Bangalore**, India for a synthetic input ground motion. They use The SHAKE2000 for the site response study. The comparison of the both test results are discussed and the correlation between corrected SPT "N" values to the G_{max} using MASW has been generated.

$$G_{max} = 11.8\,(N_{60})^{0.68} \tag{1}$$

They concluded that peak ground acceleration, maximum spectral acceleration and amplification ratio using SPT data is higher than the MASW data

The correlation between standard penetration test N values and shear modulus of low strain was examined in the suggested work in [18]. They worked to develop the relationships between SPT N values & Low Strain Shear Modulus (G_{max}).

The methods [18] and [19] investigated for Indo-Gangetic basin in Lucknow in the identification of seismic site and correlation between velocity of shear wave and standard penetration test N value. The experiments are performed to measure the seismic site classification and subsurface lithology of

the urban center of Lucknow based on borehole analysis (using SPT) and details on shallow subsurface investigations (using MASW) that are found in the Indo-Gangetic Basin (IGB) central part.
For the Indo-Gangetic basin, a correlation between SWV and N-are developed.

$Vs = 106.63\ N^{0.39}$ $R^2 = 0.47$ For Clayey soil (2)

$Vs = 60.17\ N^{0.56}$ $R^2 = 0.86$ For Sandy soil (3)

$Vs = 68.96\ N^{0.51}$ $R^2 = 0.85$ For All types of soil (4)

They concluded that:
Lithology is depicted to illustrate silty to poorly graded sand and low- to medium compressibility clay, available to a depth of approximately 30 m, using drilled borehole profiles for the subsurface. [19], [20] and [21] they work on the project, Rock depth has an impact on how seismic sites in shallow bedrock regions are classified. According to the IBC and NEHRP, these models investigate the impact of the depth of rock on the classification of zones.
They summed up their work:
It observed that a site system classification on the top 30 meters of soil layers symbolizes classes of the stiffer site having depths of shallow rock less than 25 meters from the surface of the soil.

- Here, the author proposed a new site system classification on the basis of the average thickness of soil up to engineering bedrock. For soil sites, it is considered better representative in regions of shallow bedrock.
- The analysis of one-dimensional shear wave considering the engineering bedrock depth to estimate the site periods, amplification factors, and response spectral ordinates distinct from those studies attained by assuming the top 30 meters of soil layers.

3. Methodology used in proposed work

In the present study the following parameters have been studied through experimental analysis

1. MASW experiment has been conducted in MMMUT Gorakhpur, situated on a fluvial river design of an extensive Indo-Gangetic plateau.
2. The investigation area comes under seismic zone IV, which is the zonation map of seismic in India (BIS 1893 (Part1)). The analysis territory fibs close to the seismically energetic Himalayan belt of India, making the mega-city additionally powerless.
3. Comparative study has been done for different parameters like Shear wave velocity (Vs_{30}), Response spectra, natural frequency, damping coefficient for soil present at MMMUT Gorakhpur.
4. Variation in Shear wave velocity Vs_{30} with increase in depth and with different irregularities present in soil layers has been studied.
5. Nepal earthquake 2015 records are applied to the developed models and use for Site classification of Gorakhpur.
6. Properties of soil, used in the analysis, has been found out from experimental results and comparison is done with IS 1893 Part 1 and other seismic code.

4 Experimental Analysis

The data that collected from the MASW experimental setup is feed using the software ParkSEIS (v. 3.0). After analysing the recorded data in ParkSEIS, we obtain Shear Wave Velocity (Vs_{30}). The obtain Shear Wave Velocity (Vs_{30}) is shown in table 1. The obtained $\mathbf{Vs_{30}}$ is used to plot Response Spectra for MMMUT Gorakhpur. For plotting Response Spectra, we need Earthquake Time History Data and Bore-log data SPT-N Values performed at MMMUT Gorakhpur. The Time History data used in this thesis is **Nepal Earthquake 2015 data** which is collected from Indian Metrological Department (IMD) Delhi, India. SPT- N value data is collected from Awas Vikas Parishad for the New Administration Building build at MMMUT campus Gorakhpur.

IOP Conf. Series: Materials Science and Engineering 1273 (2023) 012005 doi:10.1088/1757-899X/1273/1/012005

Table 1: Vs_{30} data for MMMUT Faculty Quarter

Layer	Depth(m)	Vs(m/s)	
1	1.163	178.88	
2	2.616	175.15	
3	4.433	178.76	
4	7.349	190.74	
5	9.542	204.91	**Vs=215m/s**
6	13.09	225.53	
7	17.526	229.97	
8	23.07	232.37	
9	30	232.94	
10	Half Space	289.87	

Table 2: Vs_{30} data for MMMUT Faculty Quarter

Layer	Depth(m)	Vs(m/s)	
1	1.163	190.57	
2	2.616	184.59	
3	4.433	140.01	
4	7.349	172.84	
5	9.542	232.24	**Vs=217m/s**
6	13.09	245.61	
7	17.526	249.88	
8	23.07	240.74	
9	30	232.55	
10	Half Space	283.91	

Table 3: Vs_{30} data for MMMUT CIVIL ENGINEERING DEPARTMENT

Layer	Depth(m)	Vs(m/s)	
1	1.163	174.24	
2	2.616	177.91	
3	4.433	152.92	
4	7.349	177	
5	9.542	253.94	**Vs=238m/s**
6	13.09	275.92	
7	17.526	276.45	
8	23.07	278.33	
9	30	274.33	
10	Half Space	336.7	

The Time History data of Nepal Earthquake 2015 is analysed using **DEEPSOIL** Software and different results such as **Ground Motion Parameters, Response Spectra, and Fourier Response spectra** are obtained. The Time History data of Nepal Earthquake 2015 and corresponding analysis is presented below.

4.1 The Time History data of Nepal Earthquake 2015

Kathmandu Valley during the 2015 Gorkha, Nepal, Earthquake. Earth Planets and Space 68:10. doi:10.1186/s40623-016-0383-7

IOP Conf. Series: Materials Science and Engineering 1273 (2023) 012005 doi:10.1088/1757-899X/1273/1/012005

Station CODE	KTP
Station Latitude.	N27.68216
Station Longitude	E85.27259
Station Height(m)	1370
Building. Type	RC 4F
Installed Floor	1F
Sampling Freq (Hz)	100
Duration Time(s)	100

First sample is at - 2015-04-25 06:11:25.95 (UTC)

5 Results and Discussions

The MASW method used for calculation of profiling of surface wave gives a quick, correct narrow and undestructive sub-surface image of the shear wave. Then the velocity model for three spots with varying soil consistence is explained. The field shear wave velocity from MASW check is in contrast with the estimated values the usage of the empirical formulae developed. It is discovered that the experimental shear wave pace and the empirically estimated values are nearly equal up to 20m depth. But at depths increased 20m, Vs from MASW check is barely greater than the calculated value the usage of empirical formulae developed.

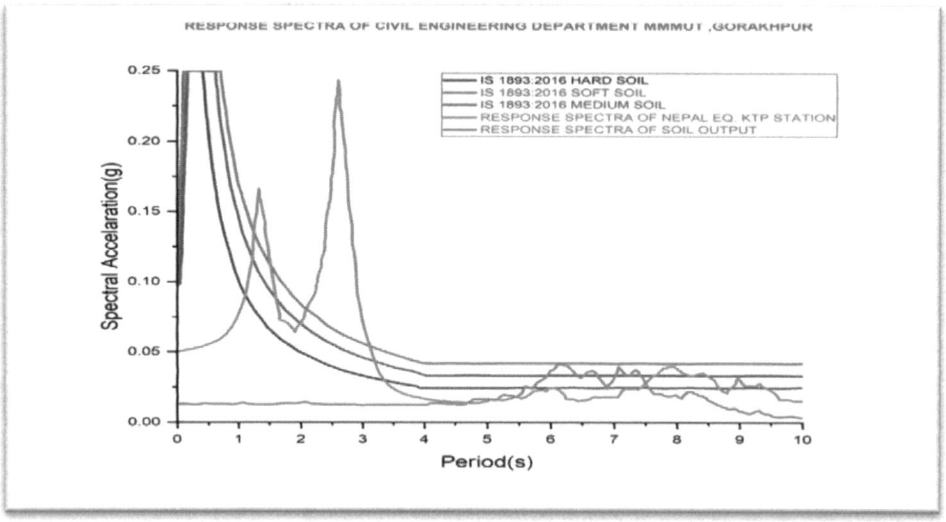

Figure 3. Response spectra of Civil engineering Department, MMMUT Gorakhpur

IOP Conf. Series: Materials Science and Engineering 1273 (2023) 012005 doi:10.1088/1757-899X/1273/1/012005

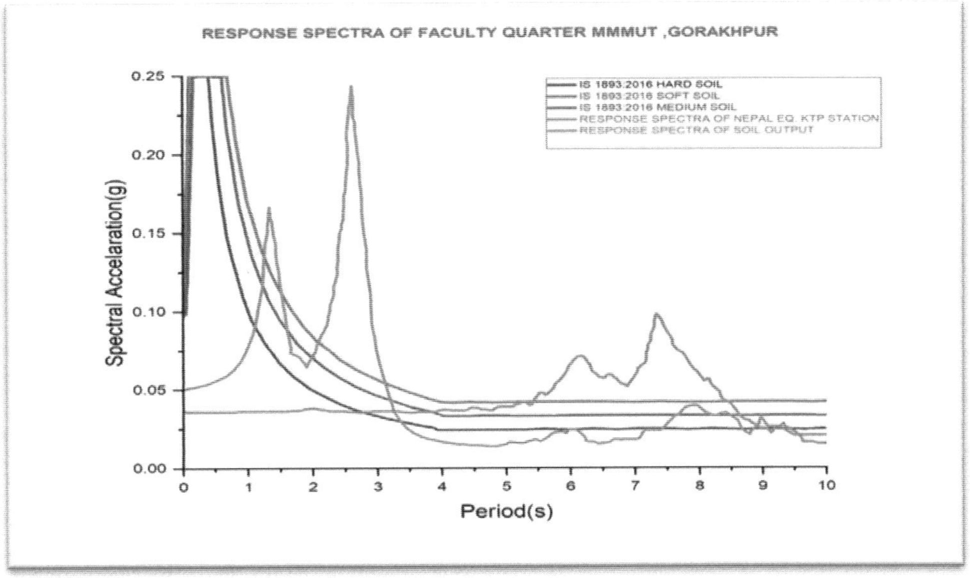

Figure 4. Response spectra of Faculty Quarter, MMMUT Gorakhpur

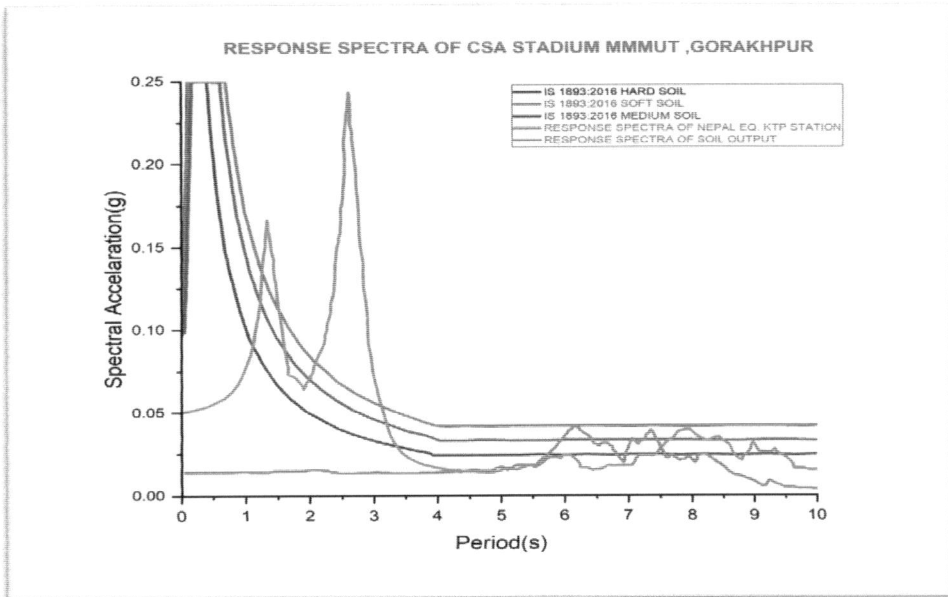

Figure 5. Response spectra of CSA Stadium, MMMUT Gorakhpur

6 Conclusion

From the graph obtained and comparing it with the **Design Response Spectra** given in **IS 1893 Part - 1 (2016)** And **Site classification as per (IBC-2009).**
Following conclusion has been found in this work:

- PGA (g) is found to be low at CSA Stadium site.
- Site near Civil Department Block 1 and CSA Stadium comes in moderate PGA (g) category for all stress drops.
- MMMUT Faculty Quarter site show maximum PGA values as high as 0.09 g.
- Output PSA (g) is showing peaks for 0 to 0.05g at all MMMUT site in addition to this, peaks were observed at 5 to 7 sec at Faculty site.
- It is recommended to plan an important structure in moderate PGA (g) areas.

- As per procedure by IBC (2009), most of sites in MMMUT Gorakhpur Campus belongs to *D*-Type and *E*-Type category.
- FS studies show that most of the areas in Gorakhpur city are liquefiable (FS < 1.0) for 0.2 g PGA, while generally MMMUT Gorakhpur Campus areas are safe at lower PGA values (0.06 g).
- As per LPI studies, liquefaction susceptibility of Gorakhpur city is 'Low' to 'Very Low' at lower PGA of 0.06 g.

7 Future Scope of this Study

- This study may serve as basis information for the more detail Site classification of Gorakhpur City.
- In this research the main site is MMMUT Gorakhpur campus, for more detail and specific characterisation of soil behaviour of Gorakhpur, more sites can be analysed using MASW technique.
- This research is an initiative of utilising modern technique of geophysical principle for soil classification of Gorakhpur City.
- As the Seismic hazard is increases day by day at the Himalayan Region, a more advance and futuristic approach for Seismic hazard management should be done, as performed in this research.
- The association between the velocity of shear wave in Gorakhpur city in the Indo-Gangetic basin & standard penetration test N value for the identification of seismic site is another focus of this research.

Acknowledgement

This paper and the research behind it would not have been possible without the exceptional support of my supervisor, Mr. Madan Chandra Maurya (Assistant Professor Civil Engineering Department, MMMUT Gorakhpur). His passion, expertise, & meticulous attention for my first experience with Response spectra to the final completion of this thesis. I'm also appreciative to my Co-Supervisor Dr. Keshav Kumar Sharma (Assistant Professor, Civil Engineering Department NIT Jamshedpur) who have also reviewed my work and patiently responded to many inquiries concerning the ideas. The kindness and knowledge of everyone have helped this research in myriad aspects and spared me from committing numerous faults; those that unavoidably persist are solely my mistakes.

References

[1] Anbazhagan, P. and Sitharam, T.G., 2010. Relationship between low strain shear modulus and standard penetration test N values. *Geotechnical Testing Journal*, *33*(2), p.1.

[2] http://www.emdat.be

[3] https://www.cred.be

[4] Anbazhagan, P. and Sitharam, T.G., 2008. Site characterization and site response studies using shear wave velocity.

[5] Atkinson, G.M. and Boore, D.M., 1995. Ground-motion relations for eastern North America. *Bulletin of the Seismological Society of America*, *85*(1), pp.17-30.

[6] Atkinson, G.M. and Boore, D.M., 2003. Empirical ground-motion relations for subduction-zone earthquakes and their application to Cascadia and other regions. *Bulletin of the Seismological Society of America*, *93*(4), pp.1703-1729.

[7] Bartlett, S.F. and Youd, T.L., 1995. Empirical prediction of liquefaction-induced lateral spread. *Journal of geotechnical engineering*, *121*(4), pp.316-329.

[8] Boominathan, A., 2004. Seismic site characterization for nuclear structures and power plants. *Current Science*, pp.1388-1397.

[9] Seed, H.B. and Idriss, I.M., 1971. Simplified procedure for evaluating soil liquefaction potential. *Journal of the Soil Mechanics and Foundations division*, *97*(9), pp.1249-1273.

[10] Seed, H.B., Wong, R.T., Idriss, I.M. and Tokimatsu, K., 1986. Moduli and damping factors for dynamic analyses of cohesion less soils. *Journal of geotechnical*

engineering, *112*(11),pp.10161032.
https://www.ucl.ac.uk/EarthSci/people/lidunka/GEOL2014/Geophysics4%20-%20waves/SEISMOLOGY%20.htm#:~:text=Seismology%20is%20the%20study%20of,are%20transmitted%20through%20the%20earth

[11] https://web.ics.purdue.edu/~braile/edumod/waves/WaveDemo.htm

[12] https://sjce.ac.in/wp-content/uploads/2018/01/EQ1-Seismicity.pdf

[13] Seismic Site Classifications and Site Amplifications for the Urban Centres in the Shallow Overburden Deposits P. Anbazhagan, M. Neaz Sheikh, DOI: 10.4018/jgee.2012010105

[14] Soil Dynamics and Earthquake Engineering. *journal homepage: www.elsevier.com/locate/soildyn*

[15] Influence of data acquisition and signal pre-processing parameters on the resolution of dispersion image from active MASW survey Jumrik Taipodia1, Arindam Dey2,4 and DipjyotiBaglari3
https://academic.oup.com/jge/articlepdf/15/4/1310/26829988/jge_15_4_1310.pdf)

[16] Earthquake Spectra and Design by N.M Newmark, W.J. Hall, Earthquake Engineering Research Institute EERI.

[17] Boore, D.M., 2004. Estimating V s (30) (or NEHRP site classes) from shallow velocity models (depths< 30 m). *Bulletin of the seismological society of America*, *94*(2), pp.591-597.

[18] Boore, D.M., Joyner, W.B. and Fumal, T.E., 1997. Equations for estimating horizontal response spectra and peak acceleration from western North American earthquakes: A summary of recent work. *Seismological research letters*, *68*(1), pp.128-153.

[19] Sonmez, H. and Gokceoglu, C., 2005. A liquefaction severity index suggested for engineering practice. *Environmental Geology*, *48*(1), pp.81-91.

[20] Kumar, V., Venkatesh, K. and Kumar, Y., 2012. Liquefaction Potential Evaluation of Alluvial Soil by Neuro Fuzzy Technique. *International Journal of Emerging Technology and Advanced Engineering, ISSN*, pp.2250-2459.

[21] Hanna, A.M., Ural, D. and Saygili, G., 2007. Evaluation of liquefaction potential of soil deposits using artificial neural networks. *Engineering Computations*.

Experimental Study on the Crack Repair Techniques of Concrete Structures: A Case Study

[1]Samreen Bano, [2]Ganesh Jaiswal, [3]Rikshit Kumar, [4]Avneesh Tiwari, [5]Karthikeyan M

[1*]Research Scholar, Department of Civil Engineering, Integral University, Lucknow, Uttar Pradesh 226026, India
[2]Assistant professor, Department of Civil Engineering, Institute of Engineering and Technology, Lucknow 226021, Uttar Pradesh, India.
[3]Assistant professor, Department of Civil Engineering, Galgotias University, Greater Noida, India
[4]Assistant professor, Department of Civil Engineering, Shri Ramswaroop Memorial College of Engineering & Management Lucknow 226010, India
[5]Assistant professor, Department of Civil Engineering, Galgotias University, Greater Noida, India
*samreenbano1az786@gmail.com

Abstract: A crack is a whole or partial split of either concrete or masonry into two or more portions caused by breaking or fracturing in concrete constructions. The bulk of fractures is caused by external forces higher than what the structure or its components can sustain acting on it. The most typical sign of degradation in concrete buildings is cracking. Once the fractured system has been assessed, an appropriate repair method that considers these reasons may be chosen. Selecting the best crack repair method may produce results that endure for a very long time while saving you a lot of time, money, and effort. The causes of cracks and several strategies for healing them are covered in this paper.

Keywords: Cracks, Epoxy Resin, CFRPs, Electrochemical methods, Bacteria

1. Introduction

Cracking is one of the most common flaws in reinforced concrete slabs and beams. The primary factors that lead to fractures include overloading, reinforcement corrosion or differential settling of the support, relatively high tensile stresses, forced deformations, plastic shrinkage, plastic settlement, and expansive reactions [7]. All of these elements have an impact on how long concrete constructions last. In recent years, buildings' upkeep and repairs have become crucial. To ensure the safety and usefulness of the structures, it is essential to repair deteriorating concrete structures. Numerous restoration techniques, including epoxy injection, route sealing, grouting, and electrochemical procedures, are available [2], depending on the nature of the cracks.

One more magnetic material for fixing is cementitious composites (ECCs), which have a broad scope of solidifying and breaking properties. It advances micro-cracks, protecting fixed structures from substances like chloride particles, water, and so on, as well as rigidity and cold execution. Notwithstanding, the conditions where these ECC benefits are accomplished should be made sense of in his solicitation [14].

Bonded overlay concrete has been a typical and generally involved fix system for the more significant part of a century with extraordinary adequacy. It has many purposes, including walkways, modern floors, and extension decks. Its allure comes from its convenience. Formwork can be restricted to unobtrusive vertical structures at the edge of the dike since it is introduced over a current substantial construction. The main aim behind a bonded overlay concrete substantial underlayment is to fix debased concrete or to develop the cross-area further and, consequently, the heap conveying limit of the construction. The reinforced liner fundamentally adds both burdens giving limit and primary unbending nature since it works solidly with the basic design. Water and dampness can't determine how to go along the surface due to the fortified layer. At last, a decent bond will want to scatter the break if the covering can't endure breaking because of unavoidable shrinkage, warm obstruction, or mechanical pressure [11].

Procedures might be chosen to achieve the following goals by carefully evaluating the reasons for cracking

- Restore and boost stiffness and strength
- Improve the durability and appearance of the concrete surface
- Improve the functional performance of the structures

The key to successfully repairing concrete cracks is to determine the source or cause of cracks, then select a recovery process or solution taking into account the reason. as well as a spacious sign of fractures in question will continue to be the "active" or "sleeping" cracks currently operating.[1]

Selection of repair methods for active and dormant cracks

The following are the best crack repair methods for active cracks

- Epoxy Injection
- Routing and sealing
- Stitching crack
- Drilling and Plugging
- Cement mortar grout [3]

2. Crack Repair Techniques in Concrete Structures

2.1 Epoxy Resin

The epoxy injection technique has been successfully used to patch fractures in concrete structures. Due to the lack of new concrete, this repair is the least expensive and invasive [8]. Sikadur 52, a low-viscosity, free-flowing, and quickly curing injection resin, is the most often used epoxy resin [5]. To expand the strength and malleability of epoxy, an adaptable relieving specialist, flexibility, and silica nanopowder were added, and two kinds of it were made: texture cover (FRP paste) and break fastener (break paste to tie materials). Epoxy tar, diluents, relieving specialist, silica nanopowder, flexibility, coupling specialist, and antifoaming specialist are utilized in FRP paste and break stick.[6]

2.2 Carbon Fibre Reinforced Polymers (CFRPs)

Using materials made of carbon fibre-reinforced polymer (CFRP), which is lightweight and robust in Spans, is beneficial when creating improvements in structural engineering. The utilization of superstructures as a retrofit approach is becoming progressively well-known. While having a great deal of knowledge of the program, using CFRPs for primary applications is gradually typical.

It has been gotten about structural designing, and considerably fewer CFRPs have been utilized to fix an assortment of designs [13]. In recent years, the Use of Carbon Fibre Reinforced Polymer (CFRP) materials in woven and laminated forms to retrofit these corroded R.C. structures. CFRP material is lightweight, corrosion-resistant, and has high tensile strength [21]. They can also quickly bond to concrete surfaces without scaffolding or large jacks and with minimal support. The CFRP system can also be used in areas where traditional procedures would be difficult to apply due to limited accessibility. As a result, CFRP repair technology makes upgrades more accessible and saves money [17].

Figure 1. Carbon Fibre Reinforced Polymers

2.3 Using Micro bacteria

The utilization of bacterially created calcium carbonate accelerates as another option and harmless to the ecosystem technique for break recuperating has been proposed. Gollapudi quickly offered this remarkable crack fix technique utilizing eco-accommodating natural cycles in 1995. A few boundaries impact microbial CaCO3 precipitation, including focus broke up inorganic carbon, pH, calcium particle fixation, and the presence of nucleation destinations. The initial three parts are given by bacterial digestion, while the bacterial cell divider fills in as the nucleation site [1].

Microbes, explicitly B. subtilis, were chosen for this examination since they meet the essential boundaries to make due in brutal conditions. Whenever Gram-positive microorganisms are presented with antagonistic conditions, they can deliver spores. This spore arrangement safeguards the life form against high mechanical pressure and antacid conditions, making it a great decision. Individuals from the class Bacillus can deliver spores that can stay lethargic for over 200 years [22].

A spectrophotometer with a frequency of 600 nm was utilized to gauge an amount of 0.5 ml of the precise arrangement. The clear was supplanted with 0.5 ml of the bacterial account after the instrument perused the clear, and the 600 nm frequency was reused [19]. Bacterial arrangements (microorganisms in a culture tank), created and handled in a controlled microbial science lab to guarantee spore creation, are regularly used to bring these microscopic organisms into processes in different substantial blends [23]

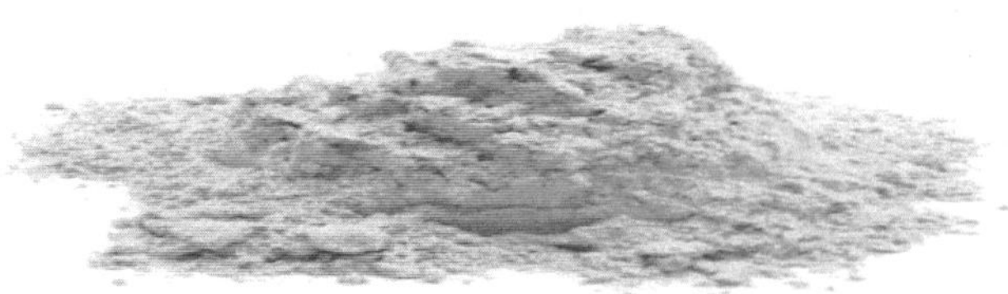

Figure 2. Bacillus subtilis spore powder

2.4 Using Electrochemical Method

The electrochemical technique is one of the handling headings to fix breaks. Examinations show it is doable to utilize electric flow to repair cracks in built-up concrete, particularly in unambiguous circumstances where other average rebuilding techniques are ineffectual. Tests are in progress to decide the practicality and advantages of utilizing electroplating innovation to fix supported substantial breaks, as well as the effect of this strategy on the mechanical and mechanical properties of the gatherings. Important primary parts [16]. Electrodeposition is another cycle used to fix supported significant breaks in marine and earthbound conditions. This methodology applies a present thickness for a brief time frame, generally half a month [18]. The utilization of current flow in the electrodeposition strategy enjoys a few upper hands over the utilization of direct flow, including lessening focus predisposition, wiping out hydrogen delicacy, diminishing blending portion, acquiring electrically stored layers that have high immaculateness and thickness, and working on the actual properties of the coating [24].

2.5 Using Bonded Overlays

Bonded overlay concrete has been a typical and generally involved fix system for the more significant part of a century with extraordinary adequacy. It has many purposes, including walkways, modern floors, and extension decks. Its allure comes from its convenience. Formwork can be restricted to unobtrusive vertical structures at the edge of the dike since it is introduced over a current substantial construction [25]. The main aim behind a bonded overlay concrete substantial underlayment is to fix debased concrete or to develop the cross-area further and, consequently, the heap conveying limit of the construction. There are various options, including those listed below, and the overlay attributes depend on the kind and quality

- Portland cement mortar and concrete
- Silica fume mortar and concrete
- Polymer concrete
- Steel fibre-reinforced polymer
- SIFCON[11]

3. CASE STUDY

"Repair of the Deira–Shindagha tunnel in Dubai" [1]-

A sea arm separates Dubai from the Arabian Gulf, and in 1975 a 561-meter-long tunnel was constructed to bridge Dubai Creek. The concrete was cast on-site with sulphate-resistant Portland cement, porous limestone, coarse aggregate, beach sand with occasional chloride fractions, and tap water. The ratio of free water to cement varied and might reach 0.6.

The concrete structure was constructed in parts with a rubber water stop at expansion joints and construction joints. Bituminized cork and a Neferma strip were utilized to fill the space between the dilation joints. The tunnel's exterior was covered with Bitu-Then sheets intended to prevent the entry of water and salt. A latex-cement (P.C.) coating was applied on the concrete inner wall with an aesthetic function.

Repair Techniques-

- Stop leakage- Epoxy injection agent was injected into the concrete around the rubber water stop in the joints. This appeared to stop the majority of the leakage effectively.
- Elimination of the impacted concrete and chloride-contaminated.This happened up to 50 mm behind the reinforcement, but it had to be limited to the reinforcement level in structurally critical sites.
- Cleaning or replacing steel reinforcement bars.

- Coating the cleaned reinforcement with an epoxy barrier coating and replacing the rebars with new rebars illustrate the cleaning and coating operation of the ramps' concrete walls.
- Replace the concrete that was removed. The initial application consisted of a polymer-modified shotcrete with blast furnace slag cement CEM III/B as the cementation's binder. Later, silica fume substituted the polymer for operational reasons.
- With the same shotcrete, the reinforcement cover was extended by 20 millimetres.
- A coating was placed that prevented oxygen from reaching the reinforcement. A 4000 m oxygen-diffusion resistance was required for the coating system. Generally, it is impossible to prevent oxygen from entering concrete during repair work. In the actual case of the underwater tunnel, however, it is conceivable if combined with additional measures.
- The coating system included two epoxy layers and a polyurethane topping. The topcoat was more resistant to ultraviolet light, which was a need for the tunnel and ramps, and could bridge cracks. Before applying the coating, the surface is sanded using an equalization slurry comprised of epoxy.

4. Experimental Studies

4.1 Epoxy Injection Technique

Six full-scale, 2.8-meter-long one-way reinforced concrete slabs were used to study how various repair methods affected the structural response of one-way slabs. Cast, cured, and tested components are 1.2 m broad and 0.2 m thick. Five high-yield deformed steel bars with a diameter of 10 mm and a characteristic strength of 460 N/mm2 make up the steel reinforcement. The concrete used has a 28-day compressive strength cube of 30 N/mm2. As illustrated in Figure 3, a two-line load is applied to the central part of each specimen during testing. Except for the control slab, which is stressed till failure, all the slabs are initially loaded to 2/3 of their projected ultimate load capacity or when the specimens begin to show indications of cracking. [5].

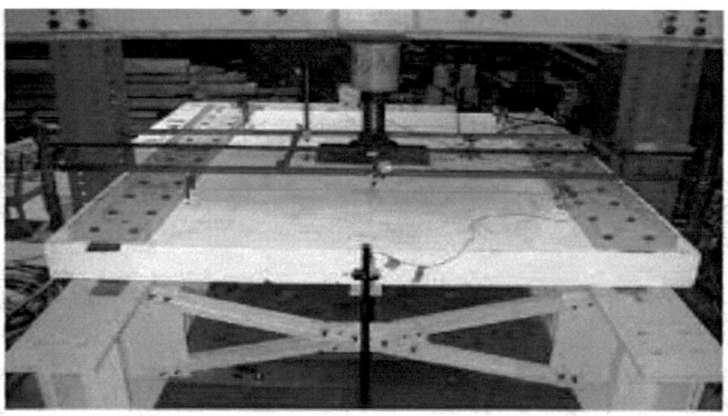

Figure 3. Test setup

Under a total load of 37.4 kN, two flexural fractures measuring 0.60 mm in width each were seen at the centre third of the slab. This restoration technique uses 200 mm-diameter injection nipples along the crack's route. The injection nipples were secured in place, and the surface of the fissures was sealed with Sikadur30. Sikadur52, a low-viscosity epoxy that is free-flowing, quick to cure, and based on a two-component solvent-free epoxy resin, is used to plug the fracture. Its properties include 1.1 kg/L density, 25 N/mm2 tensile strength after seven days, and 40 MPa compressive strength after 24 hours at 20 C. Sikadur52 puts width limitations on cracks between 0.2 and 5 mm. The crack was filled with Sikadur52 by injecting it from one end until it began to seep out of the other nip. Epoxy resin was injected into the cracks until they were filled.

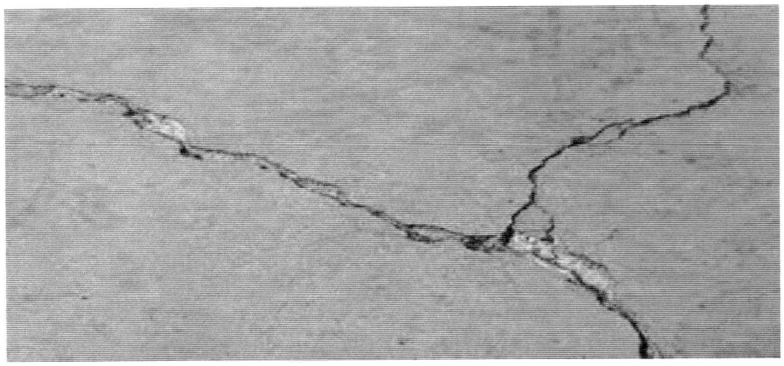

Figure 4. Cracks in slab

In the beam, to make the test simpler, concrete mortar tests are ready. In the wake of eliminating the cast, the specimens are lowered for about three months to fix. After cutting the middle part, they are relieved at room temperature for one more month. The width of the counterfeit break is set to 1 mm, and epoxy pitch is infused into the break. To fix the epoxy tar, it is held at 20 degrees Celsius for 24 hours and then at 50 degrees Celsius for 24 hours [7].

4.2 Carbon Fibre Reinforced Polymers (CFRPs)

A beam with a cross-section of 150x300mm and a length of 3350 mm was manufactured for testing. The beams are supported and loaded in flexural loading under two points loading circumstances after 28 days of cure. The fractures in unrepaired beams were tested. [15]. Three half-scale models of the inner floor column connections are exposed to a rising lateral load routine daily while being subjected to the ground load of the floor slab and beam, equivalent to dead load + 30% live load. Damaged specimens are fixed to the plate surface with an epoxy-based crack sealant and gravity-fed CFRP sheet and then subjected to the same loading technique [20].

Figure 5. CFRPs in beams

4.3 Use of Electrochemical Method

Four rebars were employed as reinforcement in the 900x150x150 mm cast beam, and the concrete's top layer was 20 mm thick. Counterfeit breaks are formed in the lowest portion of the shaft and the pillar's focal point. With break widths of 0.4, 0.6, 0.8, and 1.0 mm, the crack depths were set at 3,7,11, and 15 mm. Examples were projected, recovered after one day, and alleviated after 27 days in a typical alleviating environment. The R.C. radiates were then subjected to the electrochemical statement. [4]

5. Results and Discussions

The initial cracking loads for slabs are shown in fig 5. The repaired specimen using epoxy injection techniques shows 35% increased cracking stresses compared to controlled slabs [5].

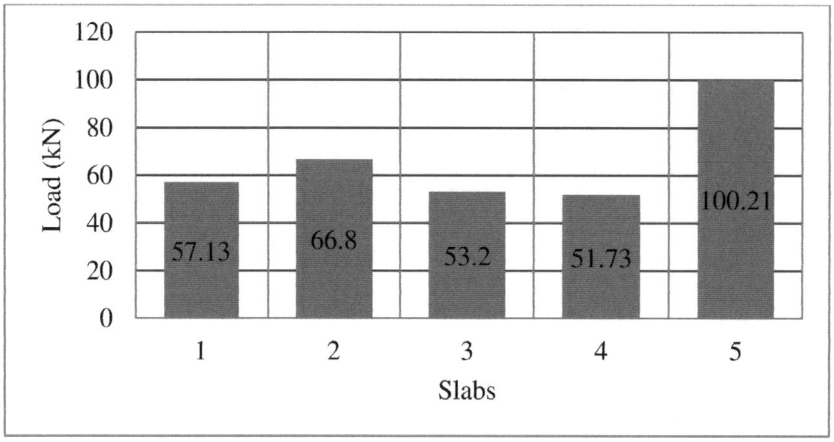

Figure 6. Cracking loads for slabs

Fig. 6 displays the control slab and various reinforced concrete slab specimens' first cracking loads (S1). All rehabilitated concrete slabs, except for S5, exhibit higher cracking loads than the control slab. Cracking loads in the restored specimens utilizing grout pouring, epoxy injection, and section expansion procedures are 35% higher than in the control slab. In comparison, adding a layer of Ferro-cement raises the cracking load by 17.8%. Reinforcing the slab by putting CFRP at its soffit does not strengthen the initial cracks; instead, it further widens the breach. Except for specimen S3, where the cracks were sealed with epoxy, all of the repair techniques used in this study can be demonstrated to be successful in restoring the damaged slab's ultimate capacity. However, the strength loss is just 8.6% compared to the control slab. The slabs S5 and S6, which were repaired using CFRP and section expansion, respectively, have ultimate load capacities of 77.4% and 130% greater than those of the control slab. The increase in maximum strength for slab specimen S5 is equivalent to that reported by Toong and Li [5], despite the comparatively low ratio of CFRP strip area to the total cross-section utilized in this experiment.

There are no notable differences in pressure debasement to 30,000 cycles for a displacement of 0.001 mm for beam fractures. Regarding the temperature swing, higher temperatures increase weakness disappointment, and the manner of disappointment in epoxy injection techniques changes from mortar to connect blow [9].

The beams are non-stationary radiation that has been put through testing up until one mm-wide fracture starts to show. This damage (due to fracture) was remedied on the specimens using CFRPs. Spokes that have undergone inspection following a repair are considered fixed. The primary break of each column seems to be 1050 mm from the focal length at the level of the second extreme. This demonstrates that support steel manufacturing begins in a nearby region. The results are made utilizing regular grades even though each layer comprises three R.C. spokes with indistinguishable cross-sectional parts and complicated supports. [12].

Then, SEM and XRD were used to look at these instances for micro-bacterial conditions. The ASTM C 39 test protocol is used to determine the compressive strength of cement. The self-recuperating limit is evaluated using a normal of five cases, whereas compressive strength is estimated using a normal of three test examples. Tetrahedrons and pyramidal shapes can be seen in bacterial cement, proving that calcite is present. It implies that the holes in the concrete mix are stuffed with calcite that has been conserved, which affects the significant's stability and toughness [10].

Following the completion of the ultrasonic measurements, samples with corrected, standardized fractures were sawed to produce cross-sections. Crystal formation at both fracture faces indicates that the untreated specimens show signs of autogenous healing during the water permeability test. After the water permeability test, no crystals were seen on the specimens with repaired fissures. Following is an explanation for this: The water flow rate for the untreated samples was so high that it took the whole test setup's top compartment to dry up between two measurements. As a result of coming into touch with the environment, the concrete's surface began to carbonate; turning Ca (OH)2 into CaCO3 crystals. Because bacteria were not resistant to the high pH in concrete, no CaCO3 crystals were seen using the microscope that was employed. However, as can be seen, the fissures are filled when the bacteria are immobilized in sol-gel. Due to the large grain size of the grout, it just covers the samples' surfaces and does not fill in the fissures.

In contrast, using epoxy led to the complete sealing of 10 mm and 20 mm deep fissures. The gel matrix cracked following sol-gel treatment alone or sol-gel+BS+CaCl2 treatment. The gel shrinks as it becomes more challenging, which causes cracking. Immediately following the filling of the cracks with silica gel and bacteria, samples treated with B.S. in sol-gel+CaCl2, Ca(NO3)2, or Ca(CH3COO)2 were put in urea-calcium solution. The fractures were filled once bacteria began to precipitate CaCO3 during immersion. However, only fissures that were 10 mm deep could be filled. The ultrasonic readings also demonstrated this. While 10 mm deep cracks treated with B.S. in sol-gel+ CaCl2, Ca(NO3)2, or Ca(CH3COO)2 worked almost as well as cracks treated with epoxy, 20 mm deep fractures were only filled when the epoxy was used.

6. Conclusion

In light of this research, one may draw the following conclusions: The rehabilitated buildings had ultimate loads and cracking that was on par with or even higher than the control slab. The flexibility and strength performances of the repaired slabs constructed using grout, epoxy injection, and Ferro-cement layers were comparable to those of the control slab. In other words, the usual reinforced concrete design for concrete slabs may be securely adopted using these restoration processes. CFRP and section expansion repair methods demonstrated increased strength-related structural performance for the broken slabs. However, compared to the control slabs, these slabs perform less ductility. One may also conclude that all forms of restoration, at the very least, successfully restore the structural integrity of fractured reinforced concrete slabs. Water permeability decreased due to the biological treatment used to seal cracks. However, it was shown that using autoclaved bacteria instead of living bacteria also caused the water flow to drop. This supports the Theory that the sol-gel matrix's capacity to patch cracks accounts for most of the decrease in water permeability. Only when active bacteria were present in the fracture healing material did a TGA study reveal the presence of CaCO3 crystals. The crystals' precipitation may improve this repair material's durability within the gel matrix. This biological therapy's effectiveness was assessed using visual inspection and ultrasonic transmission measures. After treating a fracture with B. sphaericus immobilized in silica gel, the ultrasonic pulse velocity increased, demonstrating the success of crack bridging. Visual examination of the cracks revealed that this treatment filled them. The employment of this biological repair technique is highly sought since the mineral precipitation brought on by microbial activity is unpolluted and organic. However, further research is required to determine how long this fracture-mending method will last.

IOP Conf. Series: Materials Science and Engineering 1273 (2023) 012006 doi:10.1088/1757-899X/1273/1/012006

References

[1]. Kim Van Tittelboom et al. (2010)," Use of bacteria to repair cracks in concrete", Cement and Concrete Research 40 (2010) 157–166
[2]. Panga Narasimha Reddy, Bode Venkata Kavyateja(2019)," Experimental Study on strength parameters of self-repairing Concrete", Annales de Chimie - Science des Matériaux Vol. 43, No. 5, October 2019, pp. 305-310 Journal homepage: http://iieta.org/journals/acsm
[3]. C.A.Issa Lebanese American University, Lebanon
[4]. Jiang Jhy Chang(2009)," Performance Evaluation of Using Electrochemical Deposition as a Repair Method for Reinforced Concrete Beams", S.L., vol.1, no.2, pp.75-93, 2009, t: https://www.researchgate.net/publication/267831285
[5]. Waleed A. Thanoon et.al(2005)," Repair and structural performance of initially cracked reinforced concrete slabs", Construction and Building Materials 19 (2005) 595-603 www.elsevier.com/locate/conbuildmat
[6]. Yue Li et.al(2016),"Experimental Study of Retrofitted Cracked Concrete with FRP and Nanomodified Epoxy Resin",DOI: 10.1061/(ASCE)MT.1943-5533.0001810. © 2016 American Society of Civil Engineers.
[7]. Hong-chol Shin et.al(2010),"An experimental study of fatigue resistance in epoxy injection for cracked mortar and concrete considering the temperature effect",Construction and Building Materials 25 (2011) 1316–1324
[8]. Faraj,N.S and Karim, F.R(2020),"Review On Crack Repair And Strengthening Techniques In Concrete Slabs",The International Journal of Engineering and Science (IJES) Volume 9 Issue 04 Series II Pages PP 35-48 ISSN (e): 2319-1813 ISSN (p): 20-24-1805
[9]. Ian Nicol Robertson et.al(2004),"Repair of Slab–Column Connections Using Epoxyand Carbon Fiber Reinforced Polymer",JOURNAL OF COMPOSITES FOR CONSTRUCTION J. Compos. Constr. 2004.8:376-383.
[10]. Kunamineni Vijay,Meena Murmu(2019),"Self-repairing of concrete cracks by using bacteria and basalt fiber", S.N. Applied Sciences 1:1344 https://doi.org/10.1007/s42452-019-1404-5
[11]. J. L. Silfwerbrand,"Bonded concrete overlays for repairing concrete structures",Swedish Cement and Concrete Research Institute and Royal Institute of Technology, Sweden.
[12]. S.Ahmad et.al(2011),"Use of polymer modified mortar in controlling cracks in reinforced concrete beams",Construction and Building Materials 27 (2012) 91–96
[13]. N.Haritos et.al(2006),"Repair and Strengthening of Reinforced Concrete Structures Using CFRPs",Advances in Structural Engineering Vol. 9 No. 1 2006
[14]. K.Rokugo et.al,"Patching repair with ECC on cracked concrete surface",Gifu University, Japan
[15]. Jian-he Xie and Ruo-lin Hu(2012),"Experimental study on rehabilitation of corrosion-damaged reinforced concrete beams with carbon fiber reinforced polymer",Construction and Building Materials 38 (2013) 708–716
[16]. J.S. Ryou and Nobuaki Otsuki(2005)," Experimental study on repair of concrete structural members by electrochemical method",Scripta Materialia 52 1123–1127
[17]. M.Z. Jumaat et.al(2006),"A Review of the Repair of Reinforced Concrete Beams",Journal of Applied Science Research, 2(6): 317-326
[18]. Hongqiang Chu et.al(2017),"Repair of concrete crack by pulse electrodeposition technique",Construction and Building Materials 148 241–248
[19]. Wasim Khaliq and Muhammad Basit Ehsan(2016),"Crack healing in concrete using various bio influenced self-healing techniques",Construction and Building Materials 102 349–357
[20]. Johnson, G., and Robertson, I. N. (2001). "Seismic repair and retrofit of reinforced concrete slab–column connections using CFRP." Rep. No.UHM/CE/01-04, Univ. of Hawaii, Honolulu.
[21]. David, E., C. Djelal and F. Buyle-Bodin(1998),"Repair and Strengthening of Reinforced Concrete Beams Using Composite Materials",2nd Int. PhD Symposium in Civil Engineering, Budapest.
[22]. H.G. Schlegel, General Microbiology, seventh ed., Cambridge University Press,1993.

[23]. H.M. Jonkers et.al (2010),"Application of bacteria as self-healing agent for the development of sustainable concrete",Ecol. Eng. 36 (2) (2010) 230–235.
[24]. G. Xiang(1989)," Theory and application of pulse plating", Tianjin Science Technology Press, 1989 (in Chinese).
[25]. Silfwerbrand, J. (2007), "Use of self-compacting concrete for bonded overlays", Proceedings of the 5th International RILEM Symposium on Self-Compacting Concrete, Ghent, Belgium, 3–5 September, 2, 779–784.
[26]. Srewil, Yaseen. (2006). Cracks of concrete and repair works & case study. 10.13140/RG.2.2.12299.95522.

The Implications of Sustainable Fine Aggregate on Self-Compacting Concrete: A Review

Asha Waliitagi[1](orcid.org/0000-0002-0896-9567), Vijaya Sarathy Rathanasalam[1] (orcid.org/0000-0002-5706-8589), Bramendra Kishore.H R[1], S.Jeeva chithambaram[2] (orcid.org/0000-0002-6242-4840)

[1]Department of Civil Engineering, ATRIA Institute of Technology, Bengaluru-560024, Karnataka, India.
[2]Department of Civil Engineering, Sarala Birla University, Ranchi, Jharkhand-835103, India.
Email Corresponding authors: asha.basavaraj@atria.edu

Abstract: Concrete that is self-compacting or self-consolidating is referred to as SCC with unique properties such as flowability, self-consolidation, and workability and it can be placed in any type of Reinforced cement concrete (RCC) work where reinforcements are congested and not able to do compaction. In this fast-moving modern era, the role of SCC is unavoidable. It is an advanced method of concreting and sustainable materials that can be utilized as an alternative to river sand or crushed stone sand which includes waste foundry sand, steel slag, stainless steel slag, and iron slag, among others. These materials are the product of industries and create dumping problems in the environment. These materials have a similar composition to fine aggregate there for can be replaced by some percentage with fine aggregate. To bring an alternative to existing river sand and to prevent sand mining In this novel paper detailed investigation has been made to review different sustainable material as fine aggregate and a critical review is done about the SCC features (fresh and hardened) produced using waste foundry sand, steel slag, stainless steel slag and iron slag respectively. In addition, mechanical properties along with microstructure studies of different sustainable fine aggregates are discussed. The optimum substitution rate for sustainable fine aggregate and its benefits are studied. All the alternative fine aggregate has made a significant impact on fresh and hardened properties. Due to their unique performance, various fine aggregates notably waste foundry sand, steel slag, stainless steel slag and iron slag can be employed in making SCC.

Keywords: Self-compacting concrete, Sustainable materials, Fine aggregate, and Workability.

1. Introduction

Concrete is the significant product in the building sector because of the excellent durability and mechanical qualities. Concrete on the other hand requires a large number of additional materials such aggregates, which accounts for 80% of mixture as a result, various studies have advocated finding alternatives to sand. [1]. Self-compacting concrete (SCC), created in response to the demand for effective durable and flowable concrete. It can justify it by satisfying the filling along with flowing and consolidating by its own weight hence, preserving the mixture uniformity. It is formulated by Prof. Okamura in 1986 in Japan. [1, 2].
SCC is a type of special concrete that has high deformability, strong Resistance to segregation, lower yield stress, moderate viscosity (required to guarantee consistent dispersion of solid type particles during transit, installation (without application of external compaction) and until finally the concrete sets [4]. SCC has to be fluid in its fresh form, viscous, and resistant enough to flow through reinforcements. These properties can be produced by including chemical admixtures and mineral-based admixtures particularly viscosity modifying agents (VMA) and similar superplasticizers. One of the main parameters that are the viscosity of the paste along with mortar determines how well the special concrete goes through barriers while remaining consistent. Concrete viscosity and cohesion ensure stability during the mixture and resistance to segregation, and these properties are connected to the Rheological properties of the mixes [5]. Concrete is one of the leading sustainable construction building materials because of its economical environmental source utilization, , inertness to incorporate recycled materials, and durability in the current period of rising ecological effects focus on construction building materials. Due to the concrete's inert nature and the abundance of waste, such as foundry trash, may be included into the concrete mixture with minimum impact on the qualities of the concrete. Concrete consumption has risen dramatically over the last few decades as the integration of buildings has expanded. River sand serves as a fine aggregate and is a key element in the creation of concrete. Due to the rising demand for concrete, river sand in river beds has been overused, which has had various detrimental effects, including a deeper river bed depth, a lowered water table, and salty intrusion into rivers. The restriction on sand extraction from the river raises the price of sand and has harmed the building industry's stability. As a result, It has become crucial to develop an alternative to river sand. Over the last several decades, considerable research has been undertaken on the use of industrial waste as a replacement source for fine aggregate. The study finding shows that introducing an appropriate material in concrete might increase both the mechanical and durability features, and this approach resulted in the creation of ecological concrete [6]. The different alternative materials are Foundry sand (FS) is a high silica-content byproduct of the metal alloy casting process.

2. Waste Foundry Sand (WFS)

The fast development in industrialization, especially in foundries, has resulted in a collection of waste foundry sand (WFS), which has been identified as a global environmental problem. Greensand is the main ingredient used in foundries, which is utilized frequently during the moulding process. However, after numerous mouldings, chilling, and recycling procedures, the sand becomes unsuitable and is deposited as WFS [7]. Global molding output is estimated to be at 105 thousand metric tons annually, with approximately 0.60 tonnes of WFS produced for every tonne of cast steel produced. India seems to be the world's second-largest casting producer after China (12.05 lakh metric tonnes). In India, there are around 5000 foundry plants [8]. WFS was chemically analyzed using XRF as shown in Fig.1 [9]. The microstructure of WFS is demonstrated in Fig. 2 using an SEM picture. WFS particles are round, rough textured, and angular, as represented [7].

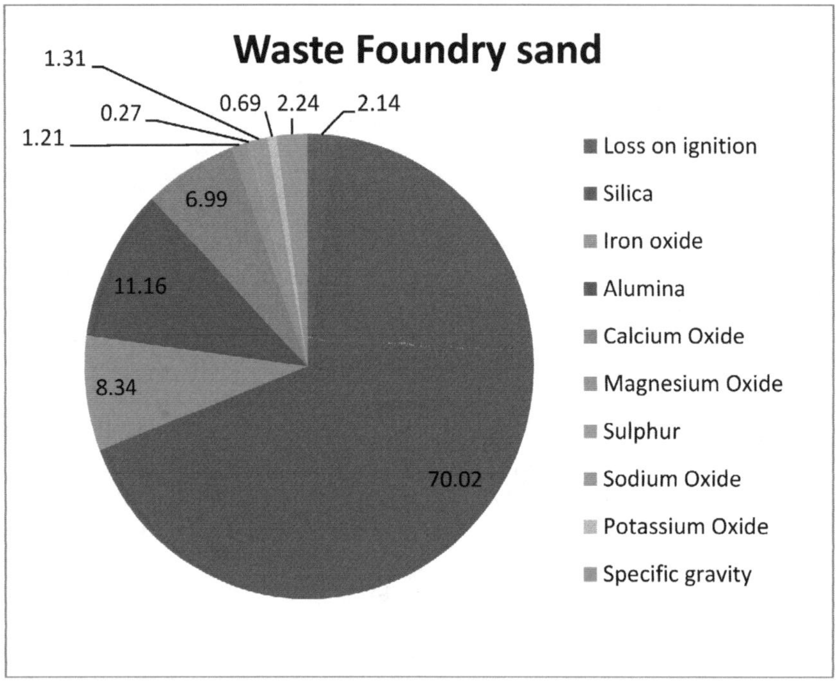

Figure 1. Chemical Assessments of Waste Foundry Sand (wt %) [9]

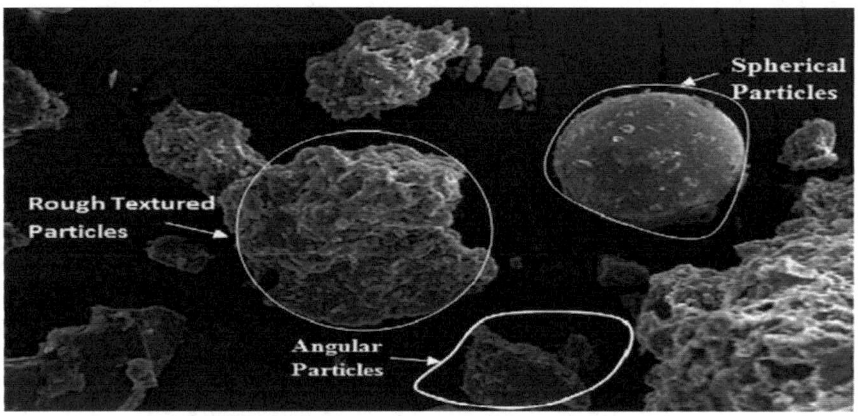

Figure 2. WFS image from an SEM [7]

2.1 Fresh properties

The work of many experts on the above-specified material was studied. The passing ability and flowability tests were executed using V- funnel, slump flow, L-box, U-box, and J-ring respectively. During the filling ability test, visual segregation was assessed. As indicated, increased WFS content resulted in a significant loss of workability. It was discovered that the colour of concrete fades to deep grey. The outcome showed that adding WFS enhanced the cohesiveness of the concrete mixtures, pointing to the possible utilization of WFS to prevent segregation. [9].

2.2 Hardened properties

The studies demonstrated with adequate strength may be obtained at an appropriate WFS replenishment ratio using natural sand. With an increasing proportion of WFS, tensile strength decreases. Tensile strength increases with age, similar to compressive strength. [7]

IOP Conf. Series: Materials Science and Engineering 1273 (2023) 012007 doi:10.1088/1757-899X/1273/1/012007

When WFS was present in concrete at 7 and 28 days the compressive strength was decreased for each of the combinations. Nevertheless, mix FS10 only offers 8.2% higher strength than the normal concrete mix after 90 days of maturity, and strength tends to decline after 10% replacement. At 7 and 28 days, the 10%, 20%, and 30% blends showed greater splitting tensile strength than the normal concrete mix [9].

According to the investigation, adding foundry sand to concrete causes a decline in compressive strength. This may be a result of the monotonic distribution of particle size of WFS, smaller nanoparticles that weaken the region between cement matrix and aggregates known as the interfacial transition zone, and inadequate packing that causes pore expansion.

3. Steel slag

Slag is a byproduct generated during the steel-making process [11]. Steel slag accounts for Slag a pebble waste product extracted from metals during ore smelting or refining. There are different kinds of slag are available like steel slag, stainless steel slag, iron slag and Electric arc furnace slag. Steel slag accounts for 12% to 20% of every tonne of steel produced. Slags are deposited in China at a rate of over 0.4 billion tonnes per year. Its usage rate, however, is less than 30%. Storing a significant volume of unusable slag consumes storage space, pollutes water, and produces other threats to the environment. It is crucial to make proper use of steel slag to solve these issues. Steel slag is a particularly valuable aggregate in concrete because of its hard, wear-resistant, adhesive, and abrasive characteristics [12, 13, and 16]. SEM image of steel slag is depicted in Fig.4.

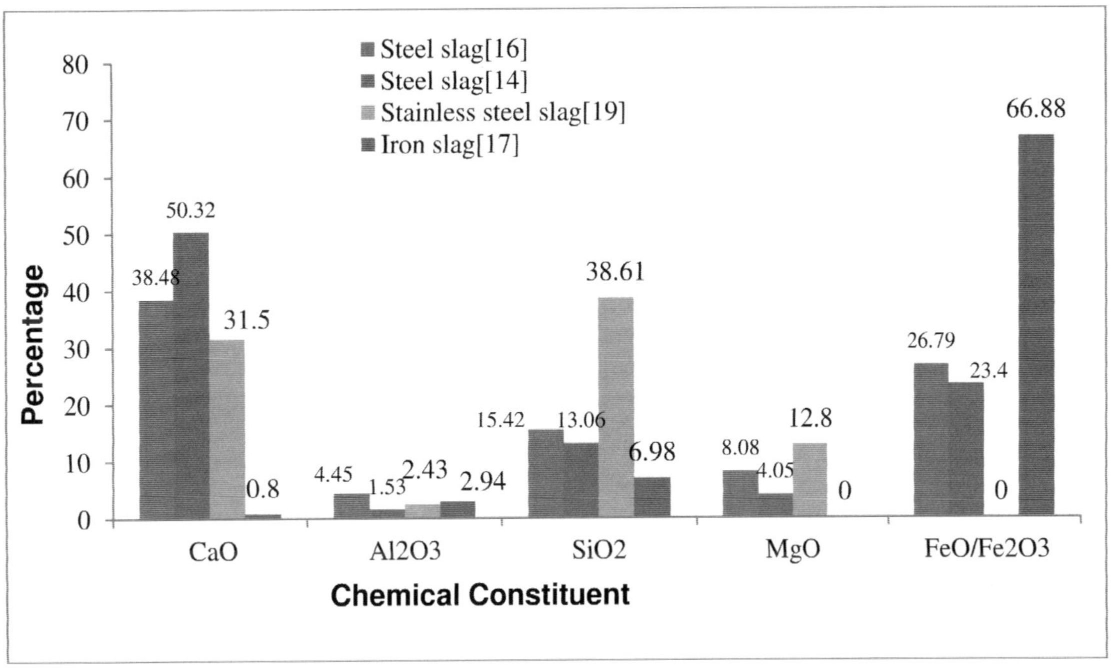

Figure 3. Chemical Formulation of Different Slag

3.1 Fresh properties

The chemical formulation of different slag is shown in Fig.3 It is represented in many types of research that substitution of normal sand with steel slag in different percentages like 20% to 100% with an interval of 20%. When 20% was replaced, there is a minor effect on the fresh mixture such as fluidity and filling performance but for 40%, 60% and 80% it was high with water-reducing agents. The result indicated that the surface contact resistance between coarse and fine aggregate increases at high steel slag to sand substitution levels, resulting in a lengthy T500 flow rate and inferior filling efficiency of the new SCC. As a result, it is fair to suggest that when assessing SCC using steel slag rather than normal sand, it should not exceed 60% [14].

3.2 Hardened properties

For all mixes, a compressive strength test was conducted. The results show that up to 60% replacement of steel slag was high compared to normal concrete cube and for 80% & 100% replacement the strength was reduced. The microstructure of steel slag shown in Figure 4 was better than normal concrete. Therefore, the optimum dosage was considered as 60% of steel slag. In another study, the strength was increased up to 40% replacement.

Figure 4. Steel Slag [14]

4 Stainless steel slag

It is a residue generated when scrap iron is used to make stainless steel. When three tonnes of stainless steel are produced, about one tonne of trash is created. This trash is problematic not only because of its volume but also because of the poisonous ingredients it contains, such as lead, cadmium, nickel and chromium which are hazardous to human and environmental health. [15, 19]. SEM image of stainless-steel slag is depicted in Fig 5.

4.1 Fresh properties

Fig.3 shows the chemical components of stainless-steel slag. The use of steel slag instead of natural sand has been demonstrated in several studies. By using the slump flow, slump-flow time (T500), V-funnel, and U Box tests, the workability of the SCC was evaluated. The properties of the materials employed in the combination, such as the shape of the particles, the distribution of their sizes, and the smoothness of their surface textures, have a major impact the workability [27]. It was discovered that the slump and slump flow were greater than those of the control mixture when the SSRS substitution ratio was up to 20%. All mixes had measured T500 times between 3 and 8 s, which is within the allowable threshold for SCC. The V-funnel phase is significantly prolonged when SSRS replacing level rises.

4.2 Hardened properties

It was visible that when more OPC was substituted out for stainless steel reducing slag (SSRS) during each testing period, lower compressive strengths were achieved, for the required 30 MPa strength at 28 days, only SCC specimens with 30% or less SSRS were acceptable. When examined at higher levels 56 and 91 days, the strength of samples with smaller than 30 per cent SSRS was somewhat lowered by around 7%, and the strength substantially fell when compared to the normal concrete [27]. When the cement's SSRS replacement ratio drops [28]. When the aggregate's stainless steel oxidizing slag (SSOS) replacement ratio decreases, the stainless steel slag's compressive strength tends to decrease rather than increase.

Figure 5. Stainless Steel Slag [15]

5 Iron slag

Around the world, substantial amounts of iron slag are produced. India makes around 17,000 tonnes of iron slag per year. A significant risk to living organisms has been posed by the large volumes of iron slag that have been deposited on the open ground nearby. The utilization of iron slag in the production of concrete is a positive step toward its eventual disposal. [17]. Durability studies of electric arc furnace slag aggregate [18] were also examined and it has a similar performance to iron slag. Performance of iron slag and steel slag [20]. Some researchers examined the performance of glass powder [21].

5.1 Fresh properties

The chemical components of iron slag are listed in Fig 3. Several types of research have proved the usage of iron slag instead of natural sand. The replacement Of iron slag in various percentages such as 10%, 25% and 40% along with this flash and superplasticizer is added. The fresh characteristics have been altered by the addition of iron slag to the SCC mixture. The workability of SCC mixtures including iron slag was shown to be decreased. The different test was performed on fresh properties like slump flow, L box, U box and V funnel. [17,22]. The slump flow values were in mm 700-750, 690-790, 640-710, 695-780, 630-790, 650-800, and 523-653 [22-26]. The observed flow values are more than 650mm therefore the passing ability was high. Similarly for the L box, U box and V funnel test the obtained values were within the limit and the flowing ability was good. With a rise in iron slag, Slump flow, L-box, and U-box values fall, but V-funnel duration increases. Because of the multi-angle and rough surface of iron slag aggregates, the workability of SCC mixes declines as the iron slag percentage increases. The results might be caused by an increase in particle friction. [17]

5.2 Hardened properties

Compressive strength testing was carried out on a 150 mm cube up to the age of 365 days in accordance with BIS: 516- 1959. It is observed that compressive strength increases with all percentages of iron slag substitution. About Control SCC, the compressive strength of SCC mixes formed with 10, 25, and 40% iron slag as fine aggregates increased by 4, 13, and 21%, respectively, after 28 days. SEM analysis revealed that when iron slag is added to concrete, the interior structure becomes denser. The SEM image shown in Fig 6a, 6b and 6c justified the reason for strength development. Utilization of sustainable materials enhances the strength properties of concrete [29, 30]

Figure 6. a. SCC at 365 days with 10 per cent of overall Iron Slag [17]

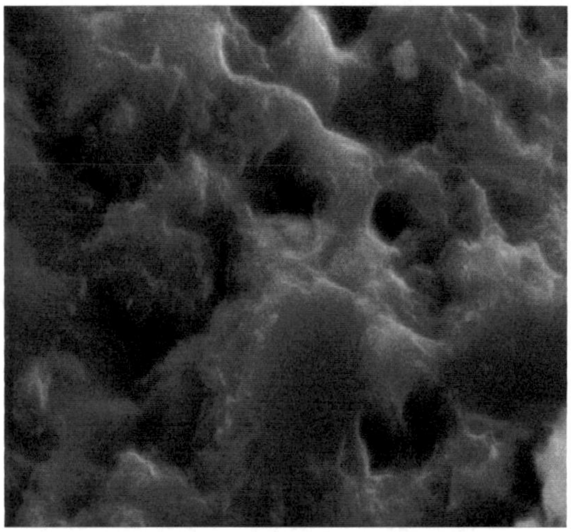

Figure 6. b. SCC at 365 days with 25 per cent of overall Iron Slag

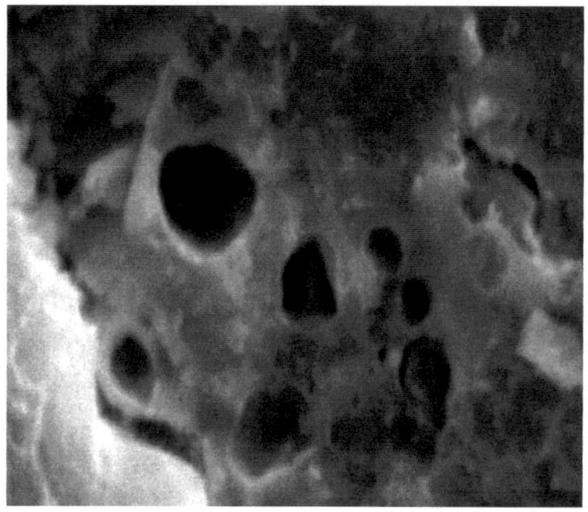

Figure 6. c. SCC at 365 days without Iro Slag [17]

6 Conclusion

According to the results observed from various research studies, the following findings are drawn

- To improve the environment, these byproducts from industry can be used in a partial dosage for the production of SCC. This would support the construction industries in terms of sustainable development.
- All these materials can be used as an alternative fine aggregate without affecting freshness, strength and durability properties.
- Irrespective of their performance Iron slag, waste foundry sand and steel slag seems to be high potential and promising and can be utilized in making SCC.
- The addition of superplasticizers along with sustainable fine aggregates enhances its fresh and mechanical properties.
- The incorporation of these materials as fine aggregate creates an eco-friendly environment by preventing sand mining.
- Replacement of these sustainable aggregates in SCC minimizes the cost of the concrete and satisfies its properties.
- Enhancement of microstructure was observed in stainless steel slag and iron slag and it may be used to generate SCC.

- The behavior of SCC with different types of materials such as weld slag, scrambled rubber and plastic waste can be investigated in further studies.

References

[1] Martins, M. A. B., Silva, L. R. R., Kuffner, B. H. B., Barros, R. M., & Melo, M. L. N. M. (2022). Behavior of high strength self-compacting concrete with marble/granite processing waste and waste foundry exhaust sand, subjected to chemical attacks. Construction and Building Materials, 323, 126492.Arjun, N., Vennila, A., & Sreevidya, V. (2017). Experimental Study on Self-Compacting Concrete with Foundry Sand and Glass Powder. *International Journal of ChemTech Research*, *10*(14), 390-395.

[2] Okamura, H., & Ouchi, M. (2003). Self-compacting concrete. *Journal of advanced concrete technology*, *1*(1), 5-15. https://doi.org/10.3151/jact.1.5

[3] Dey, S., Kumar, V. V., Goud, K. R., & Basha, S. K. J. (2021). State of the art review on self-compacting concrete using mineral admixtures. *Journal of Building Pathology and Rehabilitation*, *6*(1), 1-23. http://dx.doi.org/10.1007/s41024-021-00110-9

[4] Martins, M. A., Barros, R. M., da Silva, L. R. R., dos Santos, V. C., Lintz, R. C., Gachet, L. A., ... & Martinez, C. B. (2022). Durability indicators of high-strength self-compacting concrete with marble and granite wastes and waste foundry exhaust sand using electrochemical tests. *Construction and Building Materials*, *317*, 125907. http://dx.doi.org/10.1016/j.conbuildmat.2021.125907

[5] Ganesh Prabhu, G., Hyun, J.H., Kim, Y.Y.: Effects of foundry sand as a fine aggregate in concrete production, http://dx.doi.org/10.1016/j.conbuildmat.2014.07.070

[6] Sandhu, R. K., & Siddique, R. (2019). Strength properties and microstructural analysis of self-compacting concrete incorporating waste foundry sand. *Construction and Building Materials*, *225*, 371-383. http://dx.doi.org/10.1016/j.conbuildmat.2019.07.216

[7] Sithole, N. T., Tsotetsi, N. T., Mashifana, T., & Sillanpää, M. (2022). Alternative cleaner production of sustainable concrete from waste foundry sand and slag. *Journal of Cleaner Production*, *336*, 130399. http://dx.doi.org/10.1016/j.jclepro.2022.130399

[8] Parashar, A., Aggarwal, P., Saini, B., Aggarwal, Y., & Bishnoi, S. (2020). Study on performance enhancement of self-compacting concrete incorporating waste foundry sand. *Construction and Building Materials*, *251*, 118875. http://dx.doi.org/10.1016/j.conbuildmat.2020.118875

[9] Kanmani, S. S. Assessment of Self-Compacting Concrete Using Foundry Sand as Partial Replacement for Fine Aggregates (2019). https://doi.org/10.5281/ZENODO.2634714

[10] Feng, P., Li, Z., Zhang, S., & Yang, J. Q. (2022). Steel slag aggregate concrete filled-in FRP tubes: Volume expansion effect and axial compressive behaviour. *Construction and Building Materials*, *318*, 125961. http://dx.doi.org/10.1016/j.conbuildmat.2021.125961

[11] Liu, Z., Zeng, H., & Wang, F. (2021). Development of high performance carbonatable concrete for steel slag valorization. *Construction and Building Materials*, *291*, 123317. https://doi.org/10.1016/j.conbuildmat.2021.123317

[12] Lai, M. H., Zou, J., Yao, B., Ho, J. C. M., Zhuang, X., & Wang, Q. (2021). Improving mechanical behavior and microstructure of concrete by using BOF steel slag aggregate. *Construction and Building Materials*, *277*, 122269. https://doi.org/10.1016/j.conbuildmat.2021.122269

[13] Pan, S., Chen, D., Chen, X., Ge, G., Su, D., & Liu, C. (2020). Experimental study on the workability and stability of steel slag self-compacting concrete. *Applied Sciences*, *10*(4), 1291. http://dx.doi.org/10.3390/app10041291

[14] Sheen, Y. N., Wang, H. Y., & Sun, T. H. (2014). Properties of green concrete containing stainless steel oxidizing slag resource materials. *Construction and Building Materials*, *50*, 22-27. http://dx.doi.org/10.1016/j.conbuildmat.2013.09.017

[15] Zhuo, K. X., Liu, G. T., Lan, X. W., Zheng, D. P., Wu, S. Q., Wu, P. Z., ... & Lin, J. X. (2022).

Fracture Behavior of Steel Slag Powder-Cement-Based Concrete with Different Steel-Slag-Powder Replacement Ratios. *Materials*, *15*(6), 2243. http://dx.doi.org/10.3390/ma15062243

[16] Singh, G., & Siddique, R. (2016). Effect of iron slag as partial replacement of fine aggregates on the durability characteristics of self-compacting concrete. *Construction and Building Materials*, *128*, 88-95. http://dx.doi.org/10.1016/j.conbuildmat.2016.10.074

[17] Sosa, I., Thomas, C., Polanco, J. A., Setién, J., Sainz-Aja, J. A., & Tamayo, P. (2022). Durability of high-performance self-compacted concrete using electric arc furnace slag aggregate and cupola slag powder. *Cement and Concrete Composites*, *127*, 104399. http://dx.doi.org/10.1016/j.cemconcomp.2021.104399

[18] Sheen, Y. N., Le, D. H., & Sun, T. H. (2015). Innovative usages of stainless steel slags in developing self-compacting concrete. *Construction and Building Materials*, *101*, 268-276. http://dx.doi.org/10.1016/j.conbuildmat.2015.10.079

[19] Vaddeboina, T., & Balguri, P. K. (2022). Effect of steel slag on the properties of self compacting concrete. *Materials Today: Proceedings*. http://dx.doi.org/10.1016/j.matpr.2022.02.645

[20] Shaik. Shajahan Dr. P. Balakrishna "Study on Properties of Self Compacting Concrete with Glass Powder and Steel Slag" Published in International Journal of Trend in Scientific Research and Development (ijtsrd), ISSN: 2456-6470, Volume-4 | Issue-1, December 2019, pp.267-270,

[21] Boukendakdji, O., Kenai, S., Kadri, E. H., & Rouis, F. (2009). Effect of slag on the rheology of fresh self-compacted concrete. *Construction and Building Materials*, *23*(7), 2593-2598. http://dx.doi.org/10.1016/j.conbuildmat.2009.02.029

[22] Tomasiello, S., & Felitti, M. (2010). EAF slag in self-compacting concretes. *Facta universitatis-series: Architecture and Civil Engineering*, *8*(1), 13-21. http://dx.doi.org/10.2298/FUACE1001013T

[23] Sideris, K. K., Tassos, C., & Chatzopoulos, A. (2015). Production of durable self-compacting concrete using ladle furnace slag (LFS) as filler material. *Procedia Engineering*, *108*, 592-597. http://dx.doi.org/10.1016/j.proeng.2015.06.184

[24] Bilim, C., Atiş, C. D., Tanyildizi, H., & Karahan, O. (2009). Predicting the compressive strength of ground granulated blast furnace slag concrete using an artificial neural network. *Advances in Engineering Software*, *40*(5), 334-340. http://dx.doi.org/10.1016/j.advengsoft.2008.05.005

[25] Afshoon, I., & Sharifi, Y. (2014). Ground copper slag as a supplementary cementing material and its influence on the fresh properties of self-consolidating concrete. *The IES Journal Part A: Civil & Structural Engineering*, *7*(4), 229-242. DOI: 10.1080/19373260.2014.945622

[26] Sheen, Y. N., Le, D. H., & Sun, T. H. (2015). Greener self-compacting concrete using stainless steel reducing slag. *Construction and Building Materials*, *82*, 341-350. http://dx.doi.org/10.1016/j.conbuildmat.2015.02.081

[27] Sheen, Y. N., Huang, L. J., Sun, T. H., & Le, D. H. (2016). Engineering properties of self-compacting concrete containing stainless steel slags. *Procedia Engineering*, *142*, 79-86. http://dx.doi.org/10.1016/j.proeng.2016.02.016

[28] Chandru, P., Natarajan, C., & Karthikeyan, J. (2018). Influence of sustainable materials in strength and durability of self-compacting concrete: a review. *Journal of Building Pathology and Rehabilitation*, *3*(1), 1-16. http://dx.doi.org/10.1007/s41024-018-0037-1

[29] Chandru, P., Karthikeyan, J., Sahu, A. K., Sharma, K., & Natarajan, C. (2021). Performance evaluation between ternary blended SCC mixes containing induction furnace slag and crushed stone as coarse aggregate. *Construction and Building Materials*, *267*, 120953. http://dx.doi.org/10.1016/j.conbuildmat.2020.120953

Effect of Base Isolation under Different Soil Conditions

Gautam Rajput[1] (0000-0001-8557-4836), Vinayak mishra[2](0000-0002-4742-6549), Singh R. P.[3] (0000-0001-5409-9024)

[1](M.Tech. Structural Eng.), Civil Engineering Department, Institute of Engineering and Technology, Lucknow
[2]Assistant Professor, Civil Engineering Department, Institute of Engineering and Technology, Lucknow
[3]Chief Engineer, Central Public Works Department, New Delhi, India
rajputg926@gmail.com

Abstract. Severe seismic activity is a cause of grave concern for all kinds of structures. In order to safeguard buildings from earthquakes we use novel techniques based on performance-based seismic design. The base isolation method is one such technique utilized to reduce the effect of large-scale vibration produced during a seismic vibration. In this paper, we analyze two models: the first G+6 building and the second G+18 building using Lead Rubber Bearing (LRB). The analysis is done using ETABS software and here we use the non-linear time history method. Using the soil intrinsic effects like flexibility and stiffness, this research considers variation in displacement, drift, and repositioning of the soil in accordance with IS code 1893:2016. A greater spectral displacement was detected in medium soil in a fixed base structure. The time period for Lead Rubber Bearing building is greater than the conventional building. Base shear is minimized in (LRB) Lead rubber bearings as compared to a conventional building.

Keywords: Base isolation, Lead rubber bearing, Story Force, Time history analysis.

1. Introduction

The earthquake causes severe losses to life and property. To counter this problem multiple researchers found that isolator gives a superior representation to earthquake design procedure. Isolators are working on a procedure that decouples the structure and foundation from seismic motion, resulting in less lateral strike on the superstructure. Many scholars have been premeditated impact of SSI in building's ability to withstand earthquakes in recent years and it has been determined that relationship between soil and buildings causes impact on energy loss and a decrease in the frequency. This paper studied about the dynamic behavior of buildings that contain isolators at the base of the structure under the hard and medium types of soil Lead Rubber Bearing (LRB) is used here and their response in earthquakes is analyzed in terms of story drift, story force, story displacement and spectral acceleration. This paper is carried out in ETAB using the time history analysis method.

2. Literature review

One of the most common ways for earthquake resistance is the base isolation system. Three earthquakes have their temporal histories examined: El Centro, Loma, and Coyote found that the separated base system minimizes base shear force and story drifts when compared to the results of typical construction. Three earthquakes were subjected to time history analysis compared the results of the base-isolated condition to those obtained from the fixed-base situation & found that the base isolation system reduces base shear & story drifts. [1].

The study demonstrates that as soil flexibility and building stiffness increase, base shear value also rises. Additionally, it was noted that in soft soil conditions, the spectral acceleration is increased, providing further proof that the soil condition affects a structure's response spectrum. The soil type-I and II are acceptable for base isolated buildings, according to the paper's conclusion. To help the designer comprehend more clearly at the early stage of design & analysis additionally proposed base isolation and common structures. [2].

On the twenty-storey building, the base isolation system consists of LRB, HDRB, and a grouping of two bearings. The natural vibration, base shear & hysteretic curve are studied responses, more than the building's base isolation in terms of base shear. Results of its base shear decreases as base isolated structure's periods increase. According to research conducted by the US Geological Survey and the University of California, in the United States, a building with high damper rubber bearing & isolated base mechanism may

absorb the majority amount of energy from the earthquake vibration. [3].
This was covered in the study along with an evaluation of a fixed base structure with LRB and a conceptual examination of isolated base structures' seismic responses. The structural system being considered for the study is a three-story reinforced concrete building with one lateral degree of freedom per floor. It was determined that the LRB approach is quite effective in controlling a structure’s lateral behavior, allowing the major parameter to greatly influence how an earthquake will affect a base-isolated structure. [4].

3 Objective

In order to compare the findings of the parameters for story displacement, story forces, and storey drift as stated in IS-1893:2016 for different base isolators and to improve base isolation building performance. To design, research, and analyse the performance of building as fixed base and lead rubber bearing, as well as to look into how buildings respond to seismic pressures.

4 Detail of Modelling

We created a fixed base building with a G+6 story in the first model, a lead rubber isolated building with a G+6 story in the second model, a fixed base building with a G+18 story in the third model, and a lead rubber isolated building with a G+18 story in the fourth model.
Specifications measured with G+6 and G+18 are given away the following tables.

4.1 Structural Configuration and Design

Steel Grade	Fe 500
Concrete Grade	M30
Height of Storey	3m
Beam dimension	250×400mm
Column dimension	250×450mm
Wall thickness	230mm
Slab thickness	150mm
Parapet wall	1m
Live load on roof	3KN/m^2
Live load on Floor Finish	1.5KN/m^2
Earthquake Zone	V
Importance Factor	1
Damping Ratio	5 %
Soil Profile	Hard and Medium soil

4.2 Building specifications for (G+18)

Steel Grade	Fe 500
Concrete Grade	M30
Height of Storey	3m
Beam dimension	300×500mm
Column dimension	400×600mm
Wall thickness	230mm
Slab thickness	150mm
Parapet wall	1m
Live load on roof	3KN/m^2
Live load on Floor Finish	1.5KN/m^2
Earthquake Zone	V
Importance Factor	1
Damping Ratio	5 %
Soil Profile	Hard and Medium soil

IOP Conf. Series: Materials Science and Engineering 1273 (2023) 012008 doi:10.1088/1757-899X/1273/1/012008

4.3 Specification of isolators for G+6

Type	LRB
Vertical stiffness (U1)	1175420.57KN/m
Linear Stiffness (U2, U3)	1176KN/m
Yield strength (Q)	35KN
Damping	0.05

4.4 Specification of isolators for G+18

Type	LRB
Vertical stiffness (U1)	5588878.20KN/m
Linear Stiffness (U2 , U3)	5585KN/m
Yield strength (Q)	207.48KN
Damping	0.10

4.5 Time history function (Bhuj, 2001)

Station: India's Bhuj
Station Owner: Indian Institute of Technology, Roorkee, India's Department.
Of Earthquake Engineering
Station coordinates: 23.2480, 69.6320
Earthquake: Aftershock at Bhuj 2001-02-03 03:04:32 UTC
Hypocentral Distance: 95.0 km

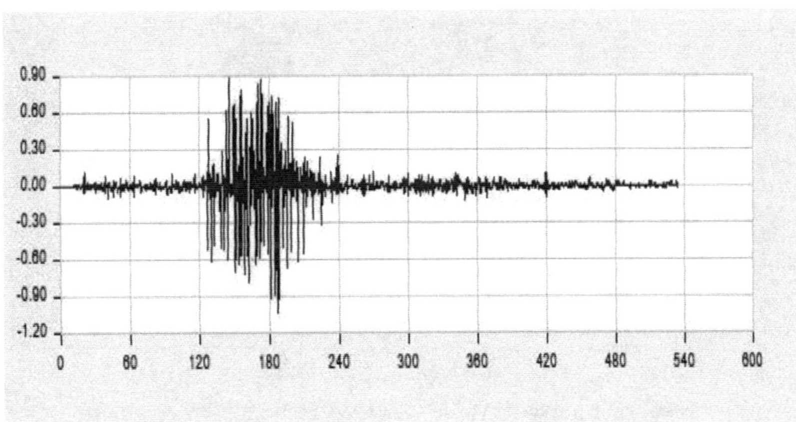

IOP Conf. Series: Materials Science and Engineering 1273 (2023) 012008 doi:10.1088/1757-899X/1273/1/012008

4.6 Model of G+6 and G+18 building

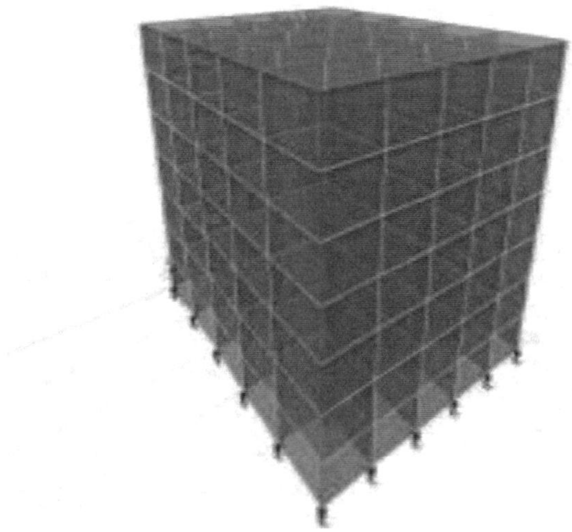

Figure 1. Model of G+6 Building

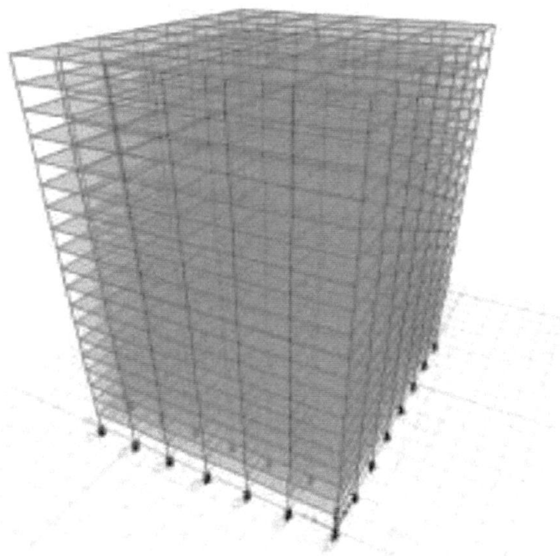

Figure 2. Model of G+18 Building

5. Results and discussion

5.1 Story displacement

The maximum displacement of Lead Rubber Bearing (LRB) and fixed base work was investigated using Time history analysis for two conditions i.e. hard and medium type of soil is taken. The results show that for LRB building at upper floor for hard type soil condition, displacement is 0.028 mm; and the fixed base is 0.031mm which is 9 % less. Furthermore, it is noted that story displacement at top story for the medium soil condition in base isolated modelis 0.038mm; and for fixed base it is 0.042mm, which is 10% less. It indicates that when the soil's stiffness increases, the story displacement decreases.

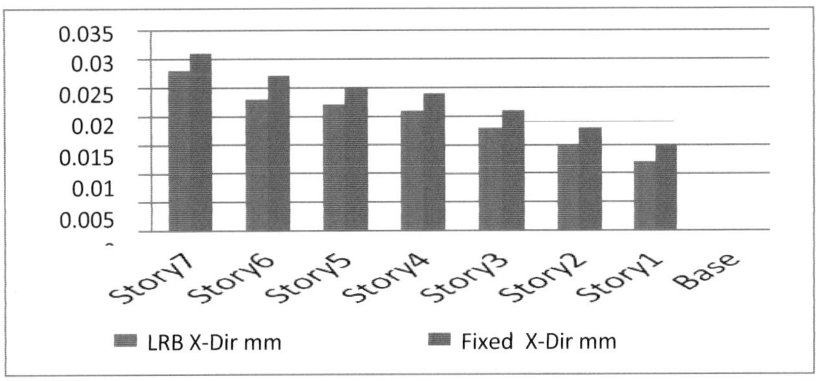

Figure 3. Story displacement, Soil-I

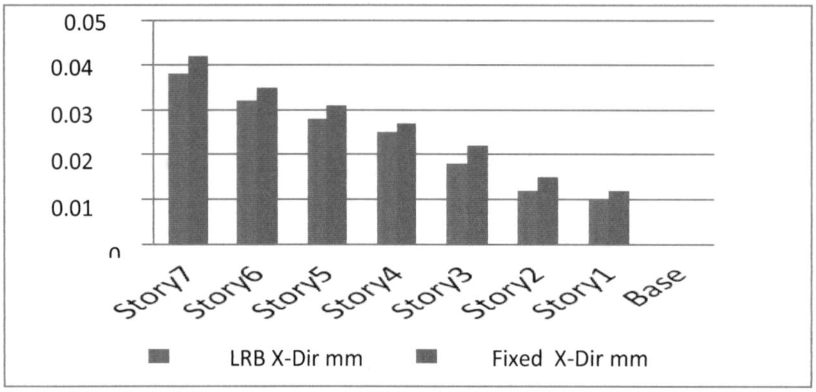

Figure 4. Story displacement, Soil-II

5.2 Story drift

The story drift is investigated using time history analysis. Drift at the first floor in G+6 buildingis 0.002256 mm and 0.002054 mm in base isolated and flexible base in hard soil condition, while in the medium soil type the drift values in base isolated and flexible base are 0.003068 mm and 0.002793 mm respectively. However in G+18 building drift value for base isolated is 0.000021 mm and for fixed base is 0.0001172 mm in hard soil condition, while in the medium soil type thedrift values in base isolated and flexible base are 0.000058 mm and 0.0001115 mm respectively. Difference of values at the first floor is 88.85% and 82.85% in hard soil and medium soil respectively.

The major difference in G+6 building between first and third story in medium soil is 0.002793 mm and 0.001366 mm with fixed base and in LRB is 0.003068 mm, 0.001707 mm, which is 9 %and 20 %, while for the hard soil is 0.002054 mm and 0.001005 mm with fixed base and in LRB is 0.002256 mm, 0.001255 mm, which is 7% and 18 % respectively. However for G+18 building in medium soil the difference between first and third story is 0.0001115 mm, 0.0002895 mm with fixed base and in LRB is 0.0000021 mm, 0.0000023 mm which is 88 % and 25 %, while forhard soil is 0.0001172 mm, 0.0000029 mm with fixed base and in LRB is 0.0000064 mm, 0.0000033 mm which is 80.85 % and 21.13 % respectively.

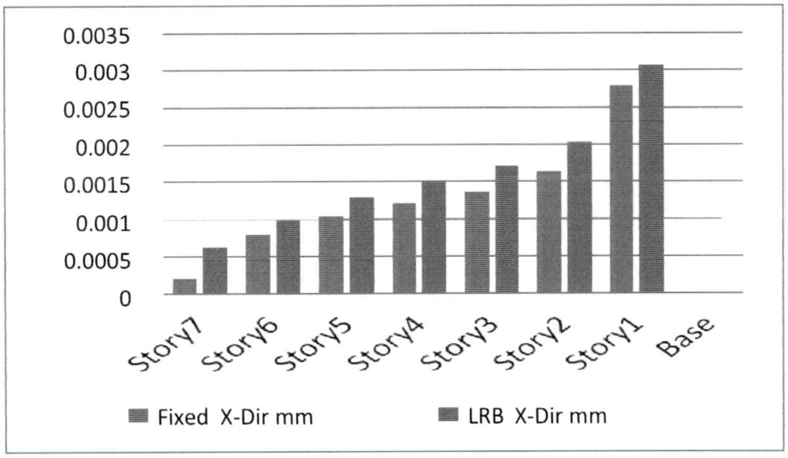

Figure 5. Story Drift, Medium soil-II (G+6)

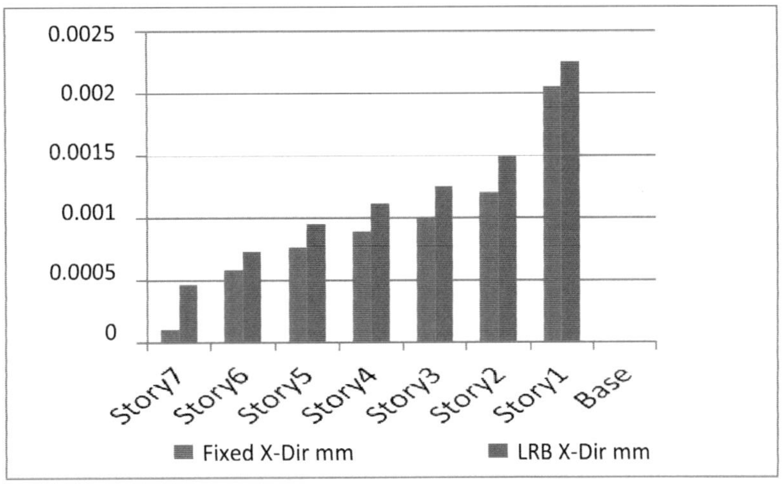

Figure 6. Story drift, Hard soil-I (G+6)

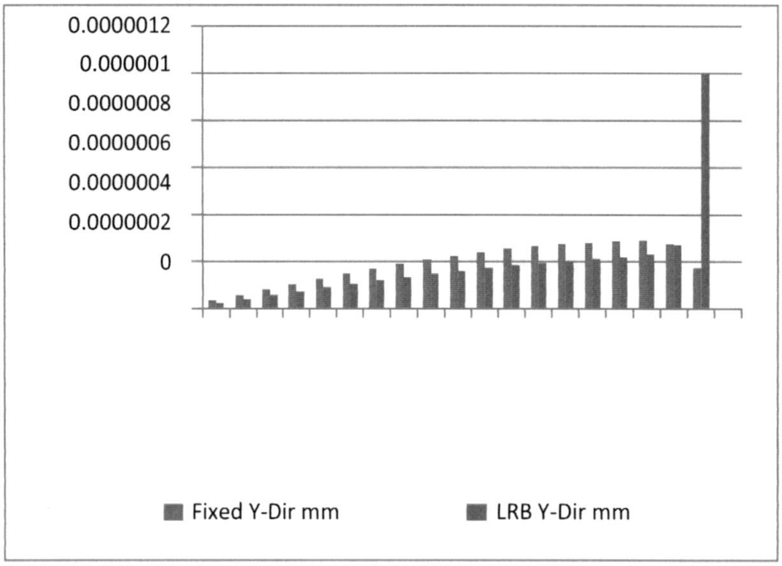

Figure 7. Story drift Medium soil-II (G+18)

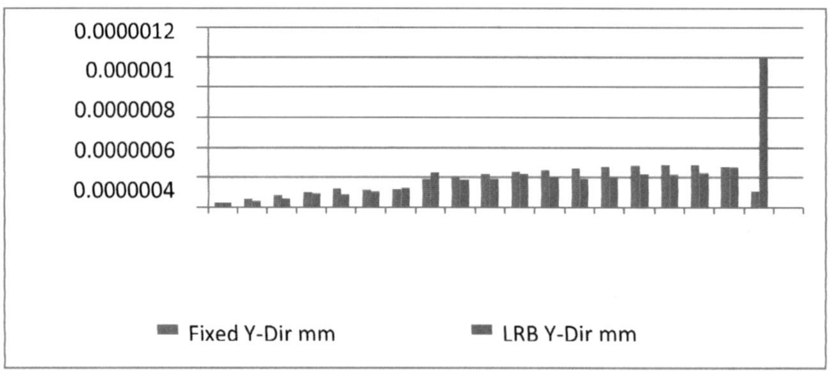

Figure 8. Story drift Hard soil-I (G+18)

IOP Conf. Series: Materials Science and Engineering 1273 (2023) 012008 doi:10.1088/1757-899X/1273/1/012008

5.3 Story Force

In case of G+6 building, the outcomes of story forces for first floor in medium soil is 178.273 kN, and 128.055 kN in fixed base & LRB respectively and in hard soil the values are 131.0831 kN, and 128.452kN in fixed base and LRB respectively. While for G+18 building, medium soil type the values of the story force are 346.541 kN and 217.401 kN in fixed base & LRB and in hard soil type, the value are 346.176 kN, and 265.665 kN/

The major difference in calculation for G+6 building is in first and sixth story is178.273kN and 14.58kN with fixed base and in isolated base is 128.055 kN and 12.4148 kN, which is 39 % and 17%. While for the hard soil is 131.0831 kN and 11.4559 kN with fixed base and in isolated base is 128.452 kN and 9.1285 kN, which is 2% and 15 % respectively. However for G+18 building in medium soil the difference between first and sixth story is 90.5775 kN and 3.01 kN with fixed base and in isolated base is 70.36 kN and 2.88 kN which is 28 % and 4 %, while for hard soil is123.18 kN and 5.66 kN with fixed base & for LRB is 101.211 kN and 3.76 kN, which is 21% and 8 % respectively. The outcomes for hard soil type with LRB outperform that of medium conditions.

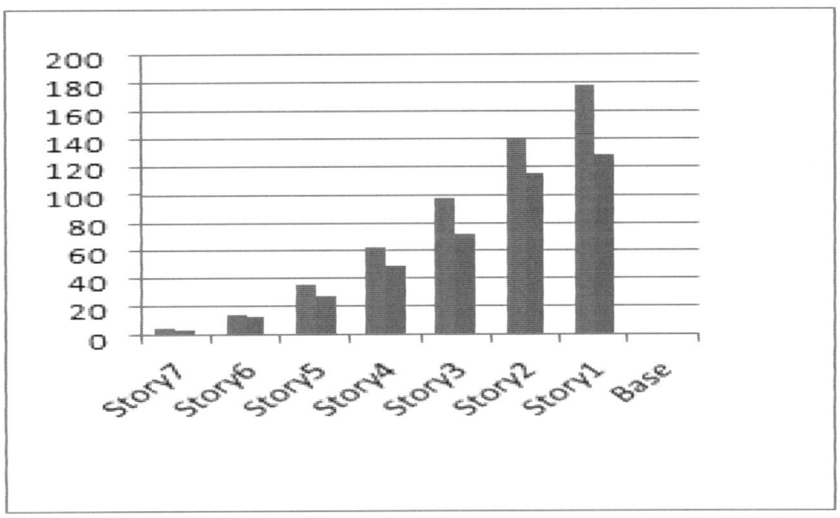

Figure 9. Story Force Medium soil-II (G+6)

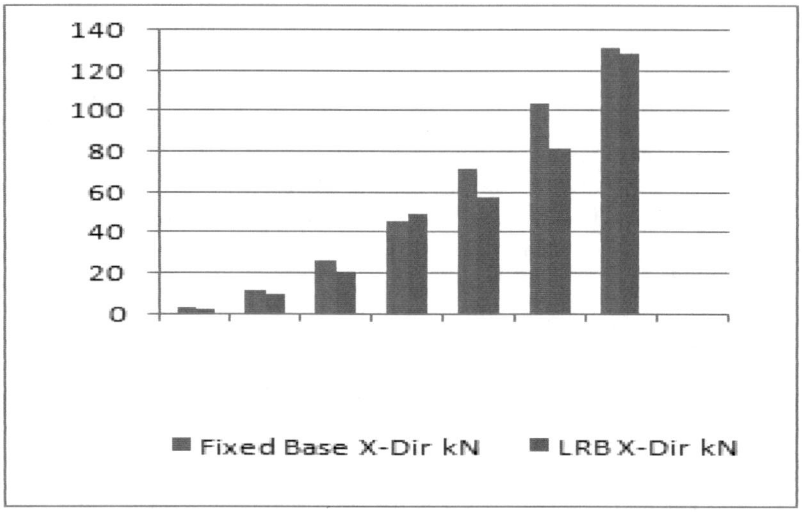

Figure 10. Story Force Hard soil-I (G+6)

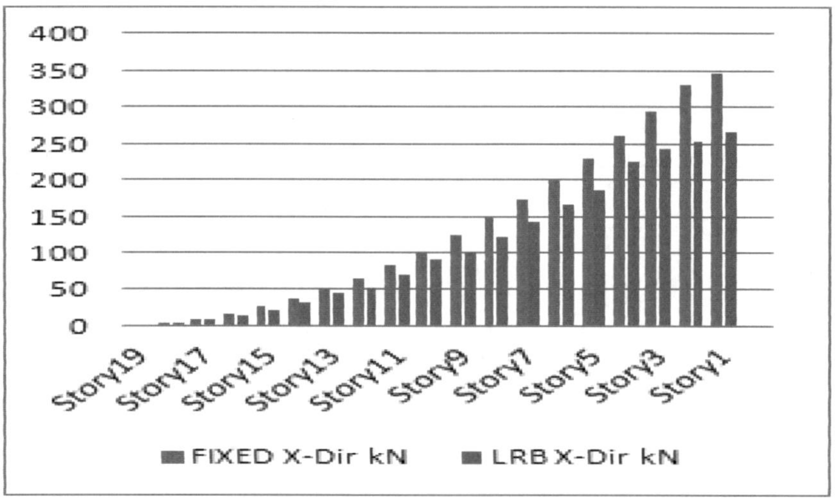

Figure 11. Story Force, Soil-II (G+18)

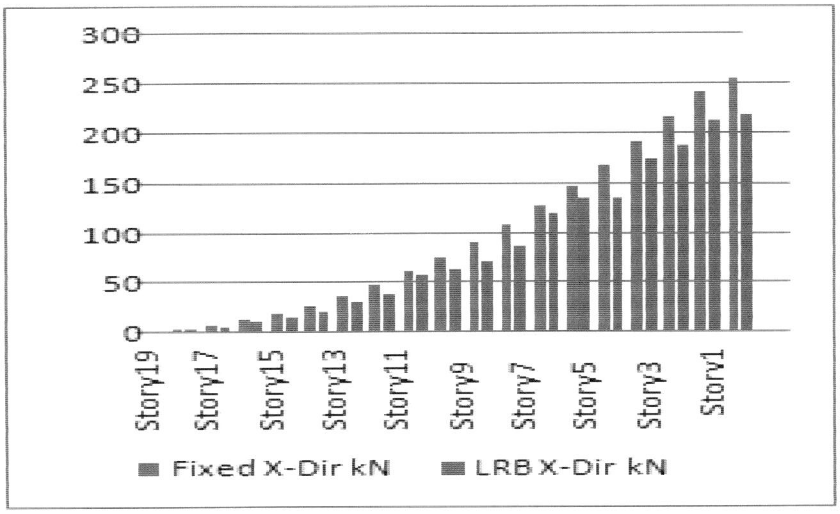

Figure 12. Story Force, soil-I (G+18)

6 Conclusion

For hard soil type and medium soil type cases, the building with LRB & fixed base was evaluated. The goal of research was to look at the seismic efficiency of LRB structure according to IS code, various soil conditions, as well as the structure's reactions, such as story drift, story displacement, and story force. Etabs software's time history analysis was used for fixed and LRB. Following study for the models, following conclusions may be drawn:

It has been noticed, the shear value increases as the soil rigidity decreases, so that this is higher for medium soil & lower for hard Soil. The paper demonstrates that story displacement rises as soil flexibility increases; to put it another way, medium Soil conditions results are more displacement as compared to hard soil conditions.

For diverse soil conditions, the maximum amount of story drift is created above the isolators on the first floor, owing to the flexibility supplied by isolators. According to the analysis of the building, a hard soil condition is acceptable for multi-story isolated buildings.

7 Future Scope of Study

Additional factors that are not considered in this analysis, such as: reinforcing bar buckling and concrete cover spalling, must be studied in order to understand the collapse/cracking mechanism of structures. A probabilistic analysis of the best base isolation strategy for multistory RC frame buildings. Research on laminated rubber bearings made from used tiers packed with

IOP Conf. Series: Materials Science and Engineering 1273 (2023) 012008 doi:10.1088/1757-899X/1273/1/012008

crushed rock to cut expenses during production. The features of materials used in isolators, such as their durability, high energy dissipation capability, large vertical stiffness, small horizontal stiffness, as well as durability can now be the subject of further research thanks to recent advancements in material technology.

References

[1] Nassani DE, Abdulmajeed MW. Seismic Base Isolation in Reinforced Concrete Structures. International Journal of Research Studies in Science, Engineering and Technology. 2015;2(2):1– 13.

[2] Hassan A, Pal S. Effect of soil condition on seismic response of isolated base buildings. International Journal of Advanced Structural Engineering. 2018 Sep 1;10(3):249–61.

[3] Farissi MA, Budiono RB. Design & analysis of base isolated structures. In 2013. p. 863–74.

[4] Tolani S, Sharma A. Effectiveness of Base Isolation Technique and Influence of Isolator Characteristics on Response of a Base Effectiveness of Base Isolation Technique and Influence of Isolator Characteristics on Response of a Base Isolated Building. (5):198–209. Available from: www.ajer.org

[5] Verma A, Pal P, Gupta YK, Allahabad I. Soil-Structure Interaction for Building Structures: A Review.

[6] Peng Y, Ma Y, Huang T, de Domenico D. Reliability-based design optimization of adaptive sliding base isolation system for improving seismic performance of structures. Reliability Engineering & System Safety. 2021 Jan;205:107167

[7] Timsina, S., & Calvi, P. M. (2021). Variable friction base isolation systems: Seismic performance and preliminary design. Journal of Earthquake Engineering, 25(1), 93-116.

[8] Jara, J. M., Hernández, E. J., & Olmos, B. A. (2021). Effect of epicentral distance on the applicability of base isolation and energy dissipation systems to improve seismic behavior of RC buildings. Engineering Structures, 230, 111727.

[9] Prajapati, M. K., & Jamle, S. (2020). Strength irregularities in multistoried building using base isolation and damper in high Seismic zone: A theoretical Review. International Journal of Advanced Engineering Research and Science, 2456-1908.

[10] Chandak N.R, "Effect of Base Isolation on the Response of Reinforced concrete Building."

Seismic response evaluation of waffle slab with and without bracing system using time history analysis

Shivani Singh[1] (0000-0001-7909-2443), Vinayak Mishra [2](0000-0002-4742-6549), and Singh R. P.[3] (0000-0001-5409-9024)

[1] M.Tech, Structural Engineering, Institute of Engineering and Technology, Lucknow
[2] Assistant Professor, Institute of Engineering and Technology, Lucknow
[3] Chief Engineer, Central Public Works, New Delhi

shivavi164@gmail.com

Abstract. In the analysis and design of structures, seismic ground motion is a significant factor. Seismic criteria consideration is necessary for serviceability and feasibility of existing as well as new structures. Time History analysis is the one in which base is subjected to a specific ground motion and dynamic response of structure is analysed at each increment of time. The bracing system increases the structure's resistance to lateral loads by transferring load from the frame to the braces. This study uses the ETABS software to do a time history analysis on a G+4 RCC building with a waffle slab and an X-type bracing system. For designing purposes, the IS 456-2000 code is thought about. According to IS 875-part 1, live loads are measured, and IS 1893-2016 earthquake analysis is carried out. For this structure, a number of factors are analysed, including storey drift, base shear, storey stiffness, and storey displacement, among others.

Keywords: Time history analysis, bracing system, time history, ETABS software, seismic response.

1 Introduction

An earthquake is a violent shaking of the Earth caused by the energy produced during the movement of tectonic plates. Earthquakes frequently cause significant harm to people and property. As a consequence of increasing urbanization and spiraling population, there is a huge demand for different types of structures while earthquakes possess severe threat to most of them [1]. Although urbanization is a progressive approach, the problems occurs when high-rise buildings are not properly designed i.e., without assessing for seismic loads. Numerous factors, including the building's shape and dimensions, earthquake intensity, and slab type, affect how well the structure performs when subjected to gravity and lateral loads [2]. In general, buildings made to support gravity loads cannot survive the horizontal vibrations of the earth. The ability of the structure to endure horizontal ground vibrations must be guaranteed. The intensity of the quakes and the qualities of the structure determine the stability and strength of a structure during an earthquake. Bracing is a cost-effective and effective method of bolstering the frame structures' resistance to lateral loads. Typically, vertically aligned areas are where the steel bracing are put [3]. Steel bracing is affordable, simple to install, occupies less space, and adds added stiffness and strength. For structures subjected to extreme lateral loads, such as earthquake loads, braced frames are very effective. Understanding how factors like storey shear, displacement, stiffness, etc. affect seismic performance requires examining variations in the seismic behavior of multi-story buildings for a variety of seismic intensities [4]. Utilizing time history analysis, the study's main objective is to assess the structural behavior of waffle slab and X-type bracing in earthquake zone IV [5].

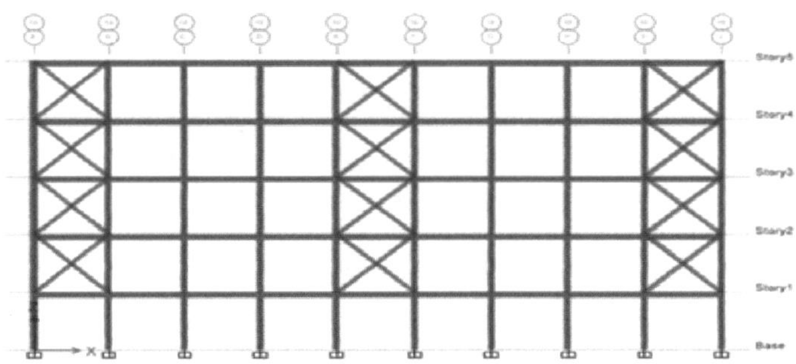

Figure 1. X- Type braced frame

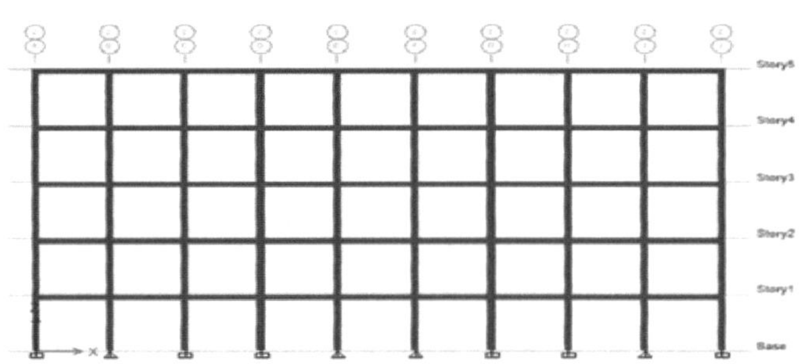

Figure 2. Non-braced frame

1.1 Waffle Slab
Waffle slabs are lighter and stiffer than solid slabs, and the voids in the waffle slabs aid to reduce the weight. Lightweight concrete is a growing trend in the construction industry as well as waffles are made to cover large span with least possible columns. It minimizes the weight of the slab by reducing the amount of concrete used [6]. They have a low dead load and a good load-caring capacity.

1.2 Chamoli Earthquake
For the time history analysis data of Chamoli earthquake has been taken into consideration. This devastating earthquake occurred on 29th March 1999 with a magnitude of 6.4 on Richter scale, struck the Himalayan seismic zone, a region known for high seismic activity. With these source parameters obtained from the literature, synthetic motion has been simulated from various sites and compared with recorded ground motion [7].

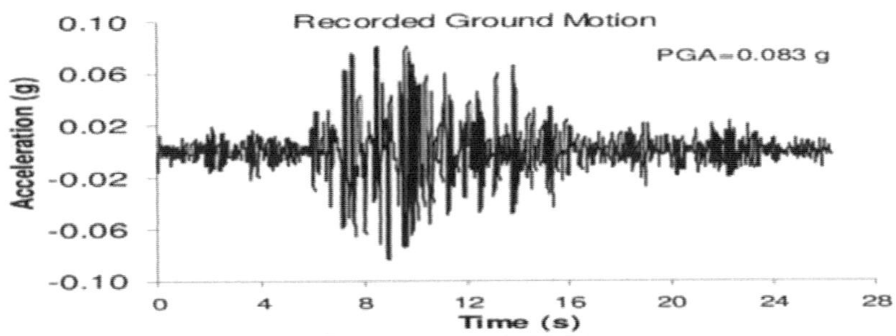

Figure 3. Typical recorded Chamoli earthquake at Ghansiali site

2 Literature Review
Y Pavan Sai Durga Reddy, Abdul Rashid, Mirza Mahaboob Baig, and others 2020: Their research's goal is to ascertain how waffle slab constructions behave when the building's hall and room have no longer any obstructing columns. This study comes to the conclusion that because self-weight increases, ribbed/waffle slab buildings are more vulnerable to lateral forces than conventional slab structures. The study was conducted in seismic zone III, and it was discovered that while ribbed slab structures have fewer columns than typical slab buildings, they perform less well in high seismic zones. However, this performance can be enhanced with structural retrofitting [8].
Pavan Kumar Jogi, Durga Chaitanya Kumar Jagarapu, and CH. Lokesh Nishanth [2020]: The goal of the project was to examine and develop a commercial building employing a variety of slab arrangements, including grid slabs, standard slabs, flat slabs with drop panels, and load-bearing walls. The major goal was to assess the financial impact of various slab layouts. They came to the conclusion that Grid slabs are safer and more cost-effective than other structural slab configurations.

Dhanaraj M. Patil, Keshav K. Sangle [2015]: The seismic response characteristics of various structures with distinct bracing methods are evaluated in order to evaluate seismic behaviour of each system [9].
Stella Mary F., Nithyambigai G., and Rameshwaran P.M. [2021]: Their investigation aims to understand waffle slab behaviour. This design of building is economical for structures with long spans of 5 to 6 metres and a gradually increasing live load. Testing a waffle slab structure with adjustable grid beam spacing's ability to carry loads was the goal of the research described in this paper [10].
The stress drops and source radius for the March 29, 1999, Chamoli earthquake were evaluated in this study, along with the source characteristics that were estimated from the strong motion data and published on the Delhi Strong Motion Accelero-graph. Y. Pandey, R. Dharmaraju, and PKS Chauhan [11].

3 Objectives

- To evaluate the seismic performance of G+4 multi-story buildings in seismic zone IV that use Waffle slabs together with and without X bracing systems.
- To assess seismic parameters for proposed structural models, such as story displacement, story drift, story shears, and story stiffness.
- To compare the seismic properties of the same structure with and without an X bracing system using time history analysis.

4 Methodology

This study uses the ETABSv16 programme to model a G+4 building by developing a 45mX45m plan and applying material properties, section properties, and load conditions. Then, using data from the Chamoli earthquake as a point of comparison, we create the time history function, and seismic zone IV is used for the study.

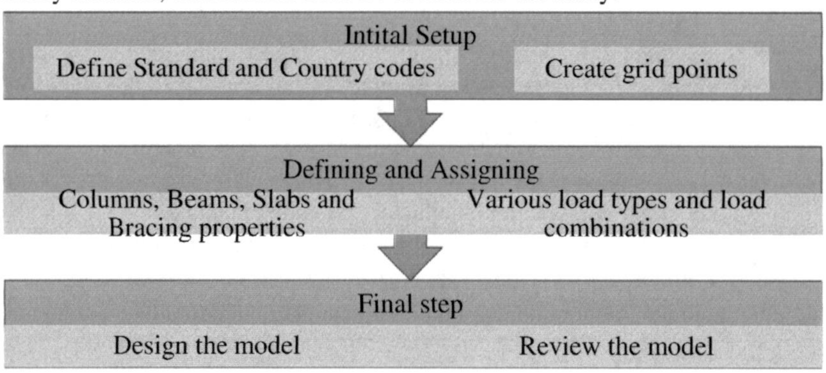

IOP Conf. Series: Materials Science and Engineering 1273 (2023) 012009 doi:10.1088/1757-899X/1273/1/012009

5 Building Parameters

5.1 *Material Properties*

Table 1. Concrete Properties

Name	Type	Design strength	E(Mpa)	Unit weight (kN/m^2)
Concrete	M30	F_c= 30 Mpa	27386.13	25
HYSD	Fe 415	F_y = 415	200000	76.9729

Table 2. Building Properties

Building Area	20.1*12.3 m2	
Building Height	18m	
Building Area	45*45 m^2	
Building Height	25m	
Columns	C1	450*450mm^2
	C2	450*350 mm^2
	C3	450*600 mm^2
Beams	B1	350*400 mm^2
	B2	350*500 mm^2
Slab Type	Waffle slab	
Slab thickness	90mm	
Spacing of ribs	1500mm	
Width of ribs	200mm	
Overall depth	800mm	
Bracings	X-type (ISA200*200*15)	

5.2 *Loading Conditions*

For dead loads, we have Here is 875 part 1, for live loads, IS 875 part 2, and seismic analysis is carried out in accordance with the 2016 edition of IS 1893 part 1 [12].

Table 3. Load Calculation

Seismic and Other Loads	Data
Live load	5 KN/m^2 (IS 875-Part 2)
Dead load	6.8 KN/m^2 (IS 875-Part 1)
Seismic zone	IV
Zone factor	0.24
Silt type	2
Acceleration due to gravity	9.81 m/sec^2

Response reduction factor	5
Importance factor	1
Scale factor (I.g/R)	1.962

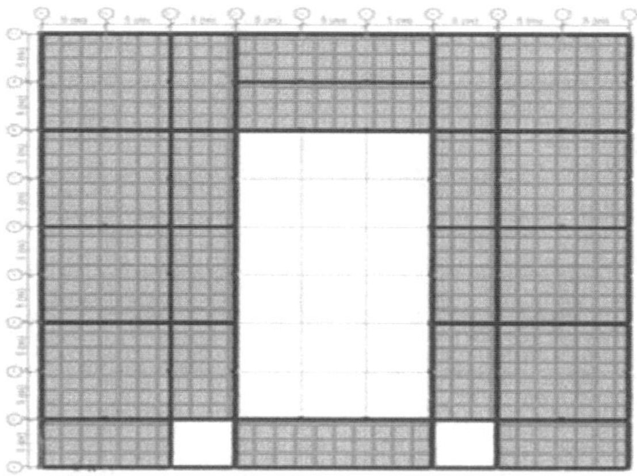

Figure 4. 2D-plan extruded plan view

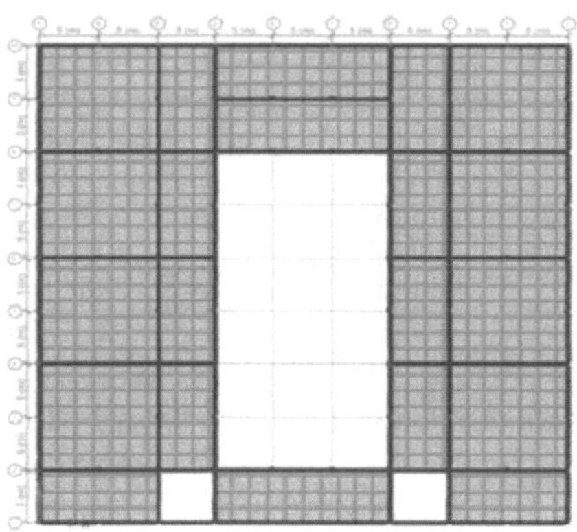

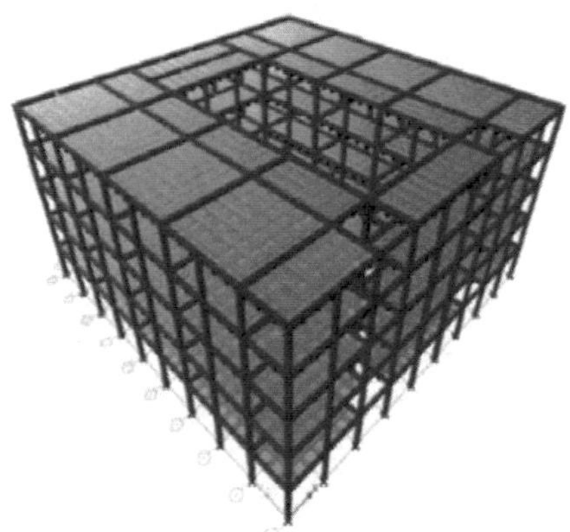

Figure 5. With bracing system **Figure 6.** Without bracing system

6 Results and Discussion

Following conclusions have been drawn from a comparison of the braced and non-braced G+4 structures:

6.1 Story Displacement

Story displacement is greatest at the top storey (12.525 mm) and lowest at the structure's base with bracing. Without bracing, the value of displacement rises together with the building's height, reaching a maximum of 27.872 mm. Displacement of model with bracing in earthquake in x direction and y direction is approx. 7.45% and 4.15% less than the maximum value of displacement without bracing. Equations are centered and set on a separate line.

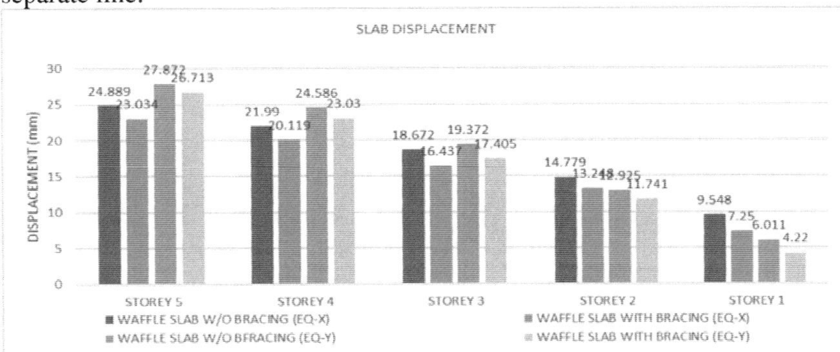

Figure 7. Story displacement

6.2 Story Stiffness

Storey Stiffness is maximum on 3rd floor and gradually decreases at the top for structure with X bracing system as compared to structure without bracing system in direction X and direction Y. (Maximum value of stiffness without bracing is 50% less than with bracing).

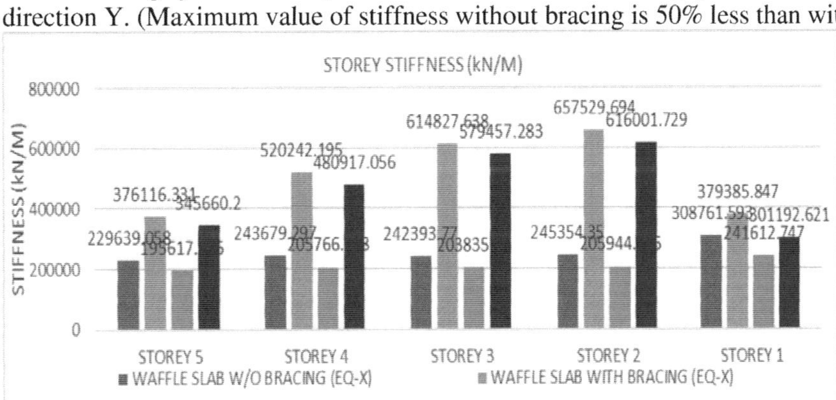

Figure 8. Story stiffness

6.3 Story Drift

According to IS 1893 (part 1) of version 2002, Cl 7.11. 3, with a part load factor of 1.0, the maximum storey drift produced by the minimum required design lateral force is restricted to 0.004 times the storey height, and based on the graphs obtained, the maximum drift is 0.001857 with bracings, which is within the permitted limits. [13].

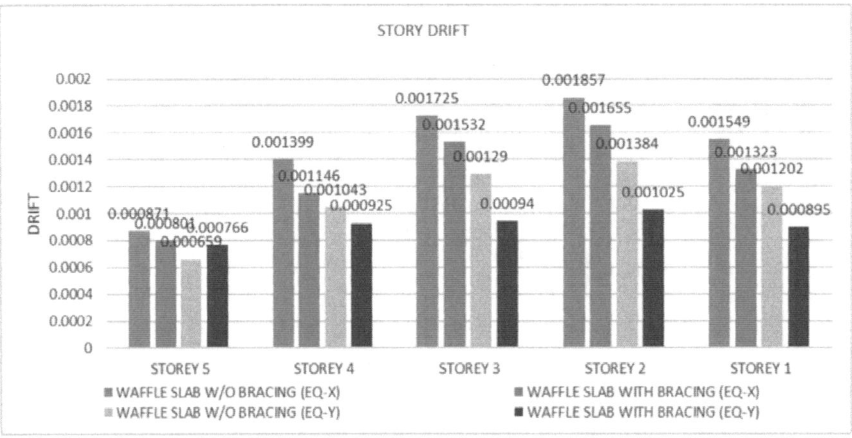

Figure 9. Story drift

6.4 *Base Shear*

The Story shear is found to be least on first floor while it increases gradually reaching the top floor. With increase of height, storey shear also increases gradually. When structure with braces and without braces compared, it is observed that storey shear shows similar trends for both.

7 Conclusion

Conclusions for the structural framework of a G+4 construction in India with different types of soil in earthquake zone IV have been established based on a time history anaysis. The time history analysis used for the study reveals significant variation in the seismic responses, including storey stiffness, storey displacements, and storey drifts in earthquake direction X as well as Y. Maximum value of stiffness for earthquake in both axes without bracing is 50% less than with bracing while displacement with bracing in earthquake in x direction and y direction is approx. 7.45% and 4.15% less than the maximum value of displacement without bracing. Maximum value of story drift for earthquake in x direction without bracing is 11% more than with bracing while story drift with bracing in earthquake in y direction approx. 25% less than the maximum value of story drift without bracing. According to design code provisions, storey drift and base shear show values that are within the acceptable ranges.

As per the results discussed the structure without bracing system is unstable and prone to damage due to which installation of bracing becomes necessary. Bracings enhance the stiffness and strength of the RC structure as well as reduce deformation by increasing the structure's stiffness and reducing displacements, drifts, etc. Waffle slabs can be adopted for large span high rise structures and exhibits good vibration control capacity.

IOP Conf. Series: Materials Science and Engineering 1273 (2023) 012009 doi:10.1088/1757-899X/1273/1/012009

8 Future Scope

- Torsion effects, joint reaction, joint displacement, and velocity may be analyzed for further discussions.
- Comparative study of different types of slabs with bracing system at different location in the structure.

References

[1] P. Anbazhagan, A. Kumar, and T. Sitharam, "Amplification from Isoseismal Map and Site Response Analysis," 3rd Indian Young Geotech. Eng. Conf. 25-26 March 20111, Indian Geotech. Soc. New Delhi, no. March 2011, pp. 1–6, 2011.

[2] "P_RJ1_004.pdf."

[3] M. M. Baig, A. Rashid, Y. P. S. D. Reddy, and T. G. N. . V. Krishna, "A comparative study on seismic analysis of G+6 building with ribbed slab & conventional slab using SAP2000 software," IOP Conf. Ser. Mater. Sci. Eng., vol. 1112, no. 1, p. 012026, 2021, doi: 10.1088/1757-899x/1112/1/012026.

[4] S. Borkar, K. Dabhekar, I. Khedikar, and N. Vaidya, "Review on 'Comparative study of flat slab structure and conventional slab structure,'" IOP Conf. Ser. Mater. Sci. Eng., vol. 1197, no. 1, p. 012072, 2021, doi: 10.1088/1757-899x/1197/1/012072.

[5] C. H. Lokesh Nishanth, Y. Sai Swaroop, D. C. K. Jagarapu, and P. K. Jogi, "Analysis and design of commercial building with different slab arrangements using ETABS," Mater. Today Proc., vol. 33, no. July, pp. 700–704, 2020, doi: 10.1016/j.matpr.2020.05.823.

[6] S. Malviya, "Comparative Study of Seismic Behaviour of Multi - Storey Buildings with Flat Slab , Waffle Slab , Ribbed Slab & Slab with Secondary Beam," vol. XIII, no. 3, 2021.

[7] G. Nithyambigai, P. M. Rameshwaran, and F. Stella Mary, "Behaviour of waffle slab," Mater. Today Proc., vol. 46, no. March, pp. 3765–3768, 2020, doi: 10.1016/j.matpr.2021.02.016.

[8] Y. Pandey, R. Dharmaraju, and P. K. S. Chauhan, "Estimation of source parameters of Chamoli Earthquake, India," Proc. Indian Acad. Sci. Earth Planet. Sci., vol. 110, no. 2, pp. 171–177, 2001, doi: 10.1007/BF02702216.

[9] D. M. Patil and K. K. Sangle, "Seismic Behaviour of Outrigger Braced Systems in High Rise 2-D Steel Buildings," Structures, vol. 8, pp. 1–16, 2016, doi: 10.1016/j.istruc.2016.07.005.

[10] A. K. Sawwalakhe and P. D. Pachpor, "Comparative Study Of Conventional Slab, Flat Slab And Grid Slab Using ETABS," IOP Conf. Ser. Mater. Sci. Eng., vol. 1197, no. 1, p. 012020, 2021, doi: 10.1088/1757-899x/1197/1/012020.

[11] Vishesh P. Thakkar, Anuj K. Chandiwala, and Unnati D. Bhagat, "Comparative Study of Seismic Behavior of Flat Slab and Conventional RC Framed Structure," Int. J. Eng. Res., vol. V6, no. 04, 2017, doi: 10.17577/ijertv6is040739.

[12] IS 456, "Plain Concrete and Reinforced," Bur. Indian Stand. Dehli, pp. 1–114, 2000.

[13] IS 1893 (Part 1), "IS 1893 (Criteria for Earthquake resistant design of structures,Part 1:General Provisions and buildings)," Bur. Indian Stand. New Delhi, vol. 1893, no. December, pp. 1–44, 2016.

IOP Conf. Series: Materials Science and Engineering 1273 (2023) 012010 doi:10.1088/1757-899X/1273/1/012010

Free Vibration Analysis of Symmetric and Asymmetric RCC Structures

***Madhuri Seeram[1](0000-0001-9504-6434), Pramod Kumar[2](0000-0002-1411-9374)**

[1] Assistant Professor, NIT Jamshedpur, Department of Civil Engineering
[2]Research Scholar, NIT Jamshedpur, Department of Civil Engineering
*Corresponding mail address: smadhuri.ce@nitjsr.ac.in

Abstract. Symmetric and asymmetric structures behave differently under external loads even though they are of an equal number of stories or equal in dimensions. It is necessary to estimate the dynamic properties of the asymmetric structures carefully to avoid failure. Modal analysis for symmetric and asymmetric multi-storied reinforced concrete (RC) buildings is very essential to assess the dynamic characteristics of the structure. The plan asymmetric buildings with the shift of core from a central position to different positions and vertical asymmetry with the mass variation in vertical direction and setback stories are considered for the present study. Flexural beam elements are modelled to represent the columns. Two translations i.e., longitudinal, and transverse displacements of the building and rotation about the vertical axis of the structure are taken as global degrees of freedom at each node. The distance between the mass center and stiffness centre on each floor is taken as eccentricity. A finite element model is developed using MATLAB code to estimate the natural frequencies. Different RCC structures of 5 to 20-storied buildings are modelled by developing global stiffness and mass matrices based on the member connectivity. Eigen-value technique is used to estimate the natural frequencies. From the study it is observed that the natural period in asymmetric structures is found more than the symmetric building. The structure with setback also showed larger natural period.

Keywords: Symmetric Building, Asymmetric Building, Centroidal position, Free Vibration, Natural frequency.

1. Introduction

Structures with asymmetric distribution of mass and stiffness in the plan or vertical direction are termed asymmetric structures. It is very difficult to construct or maintain symmetry in a building as dissimilarity may occur due to shifting of the mass centre. Eccentric rigidity due to improper design may cause asymmetry. Sometimes, encountering forces may also result in eccentricity. Rarely asymmetry in a structure may occur due to one or multiple reasons. Eccentricity in a building can be natural or accidental due to which natural or accidental torsion occurs in the building under external loading. Commonly asymmetric structures are of two types, i.e., plan asymmetry (exists due to asymmetric distribution of mass, forces, stiffness, and strength in the plan of the structure) and vertical asymmetry (due to the sudden change in mass and stiffness along the vertical direction of the structure). The soft-storey stepped and set-back buildings are a type of vertical asymmetric building. In the earlier studies, it was found that the structure behaviour depends upon the symmetric or asymmetric nature under external vibrations.

Llera and Chopra [1] developed the concept of a single Super Element (SE) to represent the stiffness properties of each floor in the multi-storied and stepped buildings considering longitudinal, and lateral displacements and rotation about the vertical axis of each floor. Based on the numerical analyses it was found that the error in the response of the building was less than 20% when compared with the actual response. They proposed that the SE model can be used for design purposes.

The accidental torsion [2] under seismic loading of asymmetric and stepped buildings was represented by the displacements in both longitudinal and lateral directions by performing a nonlinear time history analysis. In another study, Llera and Chopra (1996) [3] discussed the inelastic torsional behaviour of asymmetric buildings under seismic loading. The authors found that the orientation of the resisting plane, stiffness and strength asymmetry, strength plan-wise distribution, number of resisting planes, and power of earthquake in orthogonal direction significantly affect the torsional response. Due to the eccentricity between the centre of mass and the center of stiffness, lateral-torsional coupling in asymmetric buildings causes torsional vibration under translational ground shaking [4]. The torsional response causes a declination in the strength of the lateral load-resisting elements. The strong decline in load-resisting elements was estimated using the hysteresis model of the material. The displacements caused by strength deterioration and ductility demand of a single-story building under an earthquake loading were estimated and concluded that the effect of strength deterioration shall be considered in the inelastic design of buildings.

A simplified shear beam model [5] was used to model the resisting elements and the factor of rotational stiffness modification was developed based on the number of stories and resisting elements. The quasistatic response of each story in the studied structures was reported.

The response of an asymmetric building was estimated by assuming a flexible floor diaphragm [6] instead of a rigid floor diaphragm [7]. The analysis of asymmetric buildings has been done under uniaxial or bi-axial earth-shaking [8]. The response of the building is also affected by the angle of the earthquake force and the angle at which maximum demand is required is called the critical seismic angle [9].

As various researchers found out from previous studies that asymmetric building is very vulnerable [11-12] and soft storey building which is a type of vertical asymmetric building is the most vulnerable [13].

To assess the dynamic behavior of the structure, it is essential to perform free vibration analysis. Hence the main objective of the present study is to perform a modal analysis of the symmetric and asymmetric and setback RCC-framed buildings and compare the natural periods with symmetric building.

2. Modelling of Symmetric and Asymmetric Buildings

The symmetric structures are symmetric in their plan, geometry as well as stiffness and strength distribution. The considered RCC buildings are modeled using beam elements by considering translational motion in longitudinal and lateral direction and rotation about vertical axis. Each floor of the building is modeled with single element. The schematic diagram of the building model with idealized single element is shown in Fig. 1.

The nodal stiffness matrix for the flexural beam model of each element is given in Eq. 1.

$$k = \begin{bmatrix} k_x & 0 & -k_{xt} \\ 0 & k_y & k_{yt} \\ -k_{xt} & k_{yt} & k_t \end{bmatrix} \tag{1}$$

Where; $k_x =$ Translational stiffness in the longitudinal direction $\left(12EI_{XX}/_{l^3}\right)$

$k_y =$ Translational stiffness in the lateral direction $\left(12EI_{yy}/_{l^3}\right)$

$k_t =$ $(k_y e_x^2 + r^2\Omega_\theta^2)$ Rotational stiffness about the vertical axis

$k_{xt} =$ $k_x e_y$ Coupled stiffness in the longitudinal direction

$k_{yt} =$ $k_y e_x$ Coupled Stiffness in the lateral direction

E Modulus of elasticity of the material
I_{XX} = Area moment of inertia of the element about the x-axis
I_{yy} = Area moment of inertia of the element about the y-axis
l Length of the element
The elemental stiffness matrix is developed according to Eq. 2.

$$k_e = \begin{bmatrix} k & -k \\ -k & k \end{bmatrix} \tag{2}$$

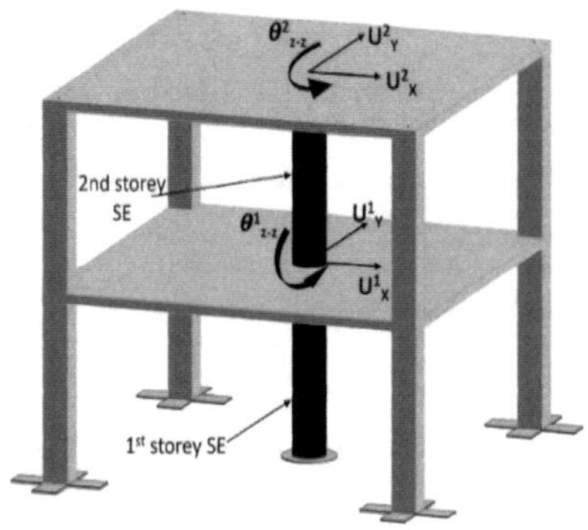

Figure 1. SE model for a building with non-aligned centers of mass

All the elemental matrices on each floor are assembled according to the connectivity of each element and formed a single element (SE). The stiffness matrix (K_{SE}) of each floor is evaluated. Single element stiffness matrices of all the floors are assembled and formed a global stiffness matrix of the building. The building's setback (if the centre of mass and centre of stiffness of two consecutive floors are not lying on a vertical axis) is considered by using a transformation matrix (Eq. 3).

$$a_u = \begin{bmatrix} 1 & 0 & 0 \\ 0 & 1 & p \\ 0 & 0 & 1 \end{bmatrix} \tag{3}$$

Where; p = $CM_g - CM_i$
CM_g the vertical centre of mass at the ground floor
CM_i the vertical centre of mass of the floor i
$b_u = (a_u)^T$
The stiffness matrix of SE for a setback floor is developed (Eq. 4) concerning the vertical centre of the mass axis of the ground floor by using the transformation technique (Eq. 4).

$$K_{SE} = \begin{bmatrix} K & -K * a_u \\ -b_u * K & b_u * K * a_u \end{bmatrix} \tag{4}$$

To develop the mass matrix, the lumped mass technique is used. Elemental lumped mass matrices are added to form a mass matrix for each floor. All the mass matrices of each floor are assembled to form the global mass matrix. The transformation technique is used in the case of setback buildings.
A MATLAB code is developed to estimate the natural period of the structures. The Eigen solution technique is used to estimate natural frequencies. The scheme of evaluation is shown in Fig. 2.

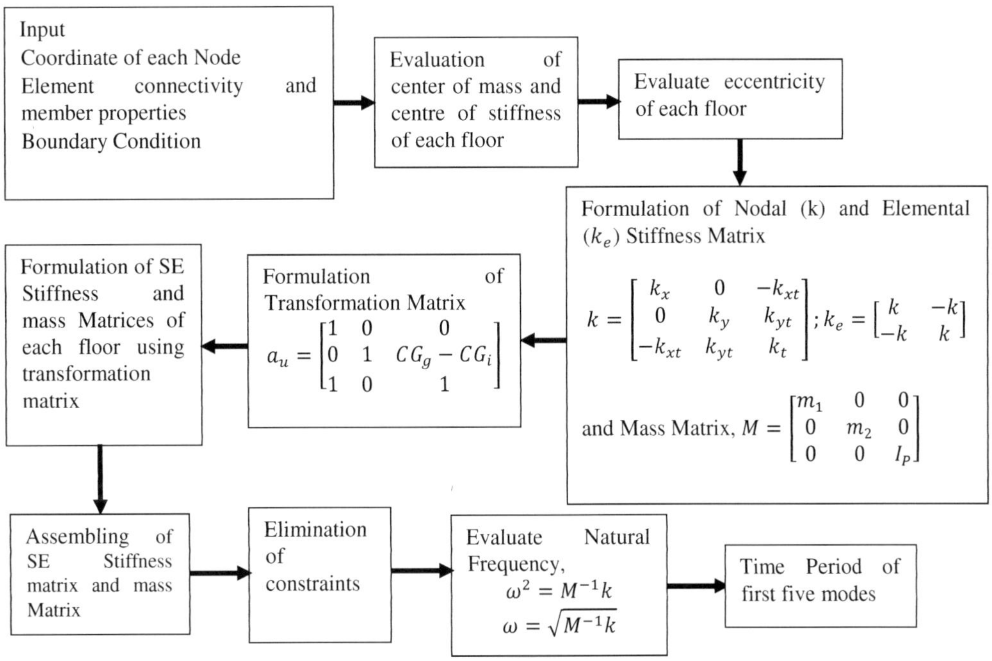

Figure 2. Flow chart of the Methodology

3. Case Studies

Symmetric, plan asymmetric and vertical asymmetric (mass and stiffness asymmetric) RCC-framed buildings have been studied. The natural period is evaluated. Free vibration analysis is performed in both cases using the developed MATLAB code. The details of the structures are discussed below.

3.1 Analysis of 10-Storey Symmetric Building

An asymmetrical ten-storied framed building with the core at the centre position is considered for the present analysis. The plan dimension of the building is 18.0m X 18.0m with four equally spaced (6m) columns of size 0.50m X 0.50m in both longitudinal and lateral directions as shown in Fig. 3. A central core of 2.0m X 2.0m with a wall thickness of 0.15m is considered a lateral force-resisting system. Each floor height is assumed as 3m and the effect of floor height is considered in the floor stiffness. A load of 8kN/m^2 is applied on each floor to simulate a dead load of beams, slab, internal walls and live load. Reinforce concrete material with a modulus of elasticity of 300kN/m^2 is assumed. In the latter case, a live load of 4 kN/m^2 is assumed on each floor but a 44 kN/m^2 additional load is applied on the top floor to model the vertical asymmetric structure.

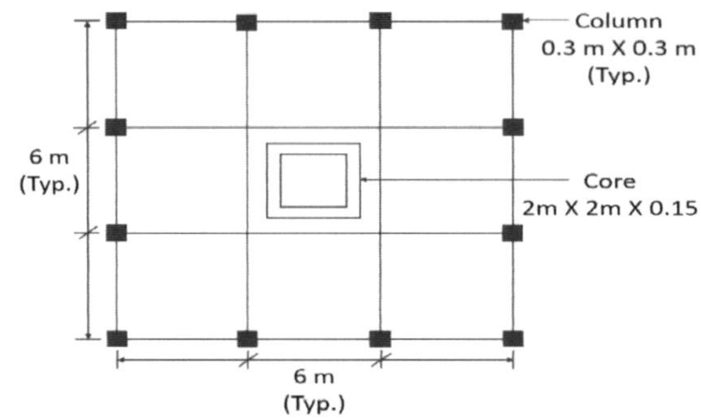

Figure 3. Plan of 10-Storied Symmetric Building (Wilkinson and Thambiratnam, 2001)

Table 1. Natural Period of 10-Storey Building

Mode	Natural period (s)	
	Symmetric	Asymmetric in vertical mass distribution
1	0.73	0.93
2	0.73	0.93
3	0.24	0.22
4	0.24	0.22
5	0.14	0.11
6	0.14	0.11

It is observed that the natural period of symmetric buildings is varied from 0.14s to 0.73s in the first six modes whereas in the vertical asymmetric case, the natural period is varied from 0.11s to 0.93s. It is also observed that the natural period is increased in the vertical asymmetric building.

3.2 Analysis of a 5-Storey Plan Asymmetric Building with a Core at a different position

A five-storied RCC plan symmetric structure is analyzed. The position of the core is changed from zero eccentricity to the maximum of 15 m as shown in Fig. 4. The size of the building in plan is about 30.0 m X 30.0 m with a typical spacing of 6.0m between the columns. The size of each column is taken as 0.3 m X 0.3 m. The core size is considered about 2.50 m X 2.50 m with a wall thickness of 0.20 m.

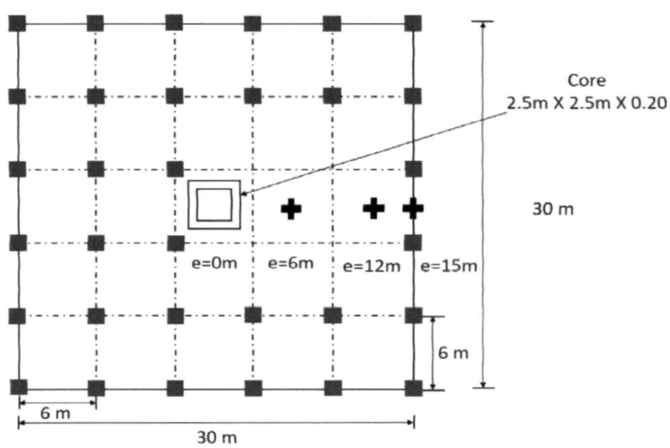

Figure 4. Plan of a 5-storey building with the core in a different position

The natural periods of five modes of the structure in the considered cases is listed in Table 2. The results show that the natural period is increasing with the increase in eccentricity. The least natural period is observed in all the five modes when the eccentricity is zero, that is symmetric building.

Table 2. Natural period for the different cases of Plan asymmetric building

Mode	Natural period (s)			
	e = 0m	e = 6m	e = 12m	e = 18m
1	0.26	1.60	2.64	2.97
2	0.16	0.50	0.82	0.92
3	0.16	0.25	0.42	0.47
4	0.08	0.16	0.21	0.24
5	0.05	0.13	0.16	0.16

3.3 Analysis of 5 and 20-Storey Plan Asymmetric Building

In this case, a plan asymmetric building has been studied. The building has five and twenty storeys (Fig. 5) with the same building plan. On each floor of the building, four different sizes of columns have been taken for the study (Table 3). The storey height of the building has been taken as 3.0m with the modulus of elasticity of concrete as 20,000 MPa. The constant mass of the floor including the mass of the beam and the mass of the floor has been taken as 360 kg/m^2.

Table 3. Details of different Column sizes

Column No.	Size of the Column (m)
1, 5, 10, 14, 15, 16	0.50 X 0.50
2, 3, 4, 13	0.46 X 0.65
6, 11	0.65 X 0.45
7, 8, 9, 12	0.60 X 0.60

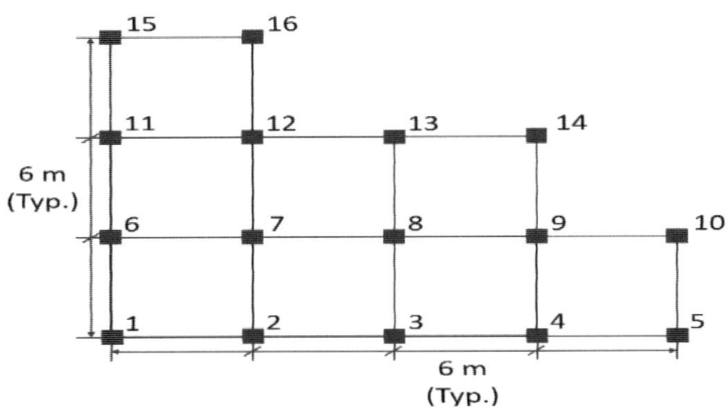

Figure 5. Plan of 5 and 20-Storey Asymmetric Building (Rafezy et al., 2007)

The natural period of the building has been evaluated and listed in Table 4. The result showed that twenty storied building has larger natural period when compared with five storied building due to the effect of slendernesss.

IOP Conf. Series: Materials Science and Engineering 1273 (2023) 012010 doi:10.1088/1757-899X/1273/1/012010

Table 4. The natural period of 5 and 20-Storey Plan Asymmetric building

Modes	Natural Period (Sec.)	
	5-Storey building	20-storey Building
1	3.84	15.99
2	3.67	15.96
3	1.19	5.31
4	1.14	5.29
5	0.61	3.16

3.4 Analysis of Vertical Asymmetric Building (Setback Building)

The present study is mainly focused on asymmetric building with setback. In modern days maximum commercial and official buildings are being constructed with setbacks for architectural views. Hence the building with setback is considered for the study. Here a three-storey setback framed building with a plan dimension of 18.0 m X 18.0 m with a storey height of 3.0 m (as shown in Fig. 6 (a) and Fig. 6 (b)) is considered. All the columns have the size of 0.30m X 0.30m. The load of beams and slab are taken as floor load of 8kPa. The effect of setback is modeled using transformation of single element of each floor. The natural period of the building has been evaluated and listed in Table 5.

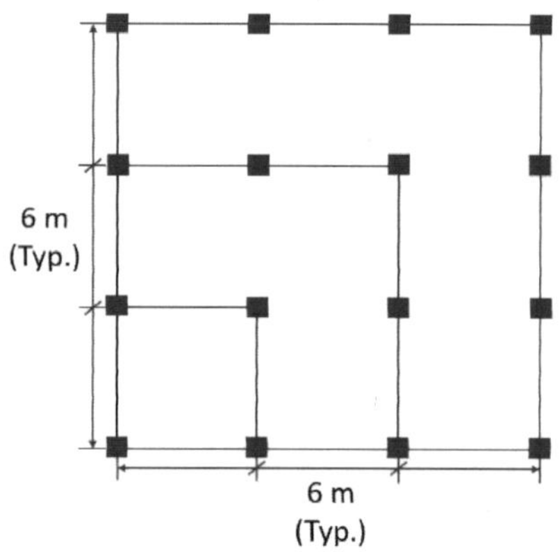

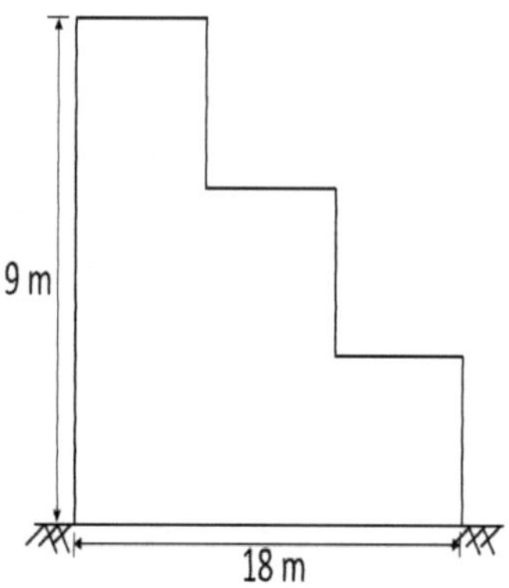

(a) (b)

Fig. 6. Three-Storey Setback Building (a) Plan (b) Elevation (Wilkinson and Thambiratnam, 2001)

Table 5. Natural Period of Setback Building

Mode	Natural period (s)
1	5.10
2	5.10
3	0.99
4	0.99
5	0.40

4 Results and Discussion

The considered RCC buildings are modeled using Single element by considering two translations and rotation about vertical axis. The symmetric and asymmetric buildings have been analyzed by developing a mathematical model and free vibration analysis is performed.

Based on the above studies, the following conclusions have been made.

1) Free vibration analysis is performed for the considered symmetric and asymmetric building models and estimated their natural period.
2) The natural period of the vertical mass asymmetric building is 27.39% more than the symmetric building (Table 1). It is observed due 11 times more mass concentrated on the top floor when compared with other floors mass of the building.
3) It is observed that as the degree of in-plan eccentricity increases in the building the natural period increases with the increase in eccentricity (Table 2). This phenomenon is observed due to the increase in moment of inertia due to eccentricity in the building.
4) In the plan asymmetric buildings, as the height of the building increases the natural period is increased (Table 4) due to the slenderness effect.
5) In the vertical asymmetric building with setback, the natural period is found more than 5s.

5 Conclusions

1) From the study, it is observed that the natural period is increased in both plan and vertical asymmetric buildings when compared with symmetric buildings.
2) The natural period also increased with the increase in the height of the building.
3) The natural period is increased as the eccentricity increases in both plan and vertical asymmetric buildings.

References

[1] Juan C. De La Llera and A. K. Chopra (1995) "A simplified model for analysis and design of asymmetric-plan building" Earthquake Engineering and Structural Dynamics, 24: 573-594. https://doi.org/10.1002/eqe.4290240408
[2] Juan C. De La Llera and A. K. Chopra (1996) "Accidental and Natural Torsion in Earthquake Response and Design of Building" Eleventh World Conference on Earthquake Engineering.
[3] Llera, J. C. D. L., & Chopra, A. K. (1996). Inelastic behaviour of asymmetric multistory buildings. Journal of structural engineering, 122(6), 597-606. https://doi.org/10.1061/(ASCE)0733-9445(1996)122:6(597)
[4] S. C. Dutta (2001) "Effect of Strength Deterioration on Inelastic Seismic Torsional Behaviour of Asymmetric RC Buildings" Building and Environment, 36: 1109-1118. https://doi.org/10.1016/S0360-1323(00)00076-7
[5] Sean Wilkinson and David Thambiratnam (2001) "Simplified Procedure for Seismic Analysis of Asymmetric Building" Computer and Structures, 79: 2833-2845. https://doi.org/10.1016/S0045-7949(01)00150-X
[6] Goel, R. K., & Chopra, A. K. (1993). Seismic code analysis of buildings without locating centres of rigidity. Journal of Structural Engineering, 119(10), 3039-3055. https://doi.org/10.1061/(ASCE)0733-9445(1993)119:10(3039)
[7] D. Basu and S. K. Jain (2004) "Seismic Analysis of Asymmetric Buildings with Flexible Floor Diaphragms" Journals of Structural Engineering, 130: 1169-1176.
[8] D. Marusic and P. Fajfar (2005) "On the inelastic seismic response of asymmetric buildings under bi-axial excitation" Earthquake Engng Struct. Dyn., 34:943-963. https://doi.org/10.1002/eqe.463
[9] Magliulo, G., Maddaloni, G., & Petrone, C. (2014). Influence of earthquake direction on the seismic response of irregular plan RC frame buildings. Earthquake Engineering and Engineering Vibration, 13(2), 243-256.
[10] B. Myslimaj and W. K. Tso (2005) "A Design-Oriented Approach to Strength Distribution in Single-Story Asymmetric System with Elements Having Strength-Dependent Stiffness" Earthquake Spectra, 21: 197-212. https://doi.org/10.1193/1.1854152
[11] B. Rafezy et al. (2007) "Coupled Lateral-Torsional Frequencies of Asymmetric, Three-Dimensional Frame Structures" International Journal of Solid and Structures, 44: 128-144. https://doi.org/10.1016/j.ijsolstr.2006.04.019

[12] Hentri, M., Hemsas, M., & Nedjar, D. (2021). Vulnerability of asymmetric multi-storey buildings in the context of performance-based seismic design. European Journal of Environmental and Civil Engineering, 25(5), 813-834.
[13] Sinha, A. K., & Bose, P. R. (2004, August). Seismic Vulnerability Assessment of Asymmetric Structures. In 13th World Conference on Earthquake Engineering, Vancouver, BC, Canada, Paper (No. 1478).
[14] Jain, S. K., Murty, C. V. R., Dayal, U., Arlekar, J. N., & Chaubey, S. K. (2001). The republic day earthquake in the land of MK Gandhi, the father of the Nation. Department of Civil Engineering, Indian Institute of Technology Kanpur,

Comparative Analysis of Multistorey RCC and Semi-Rigid Steel Frame with Cross Bracings Subjected toMainshock and Aftershock Earthquake Sequence

Sharad Srivastava[1] (0000-0001-9186-5971) and Dinesh Narain Srivastava[2] (0000-0002-0392-2209)

[1] M.Tech, Structural Engineering, Institute of Engineering and Technology, Lucknow

[2] Executive Engineer, Uttar Pradesh Rajya Nirman Sahkari Sangh Limited, Lucknow

Corresponding email: sharad.sri011@gmail.com

Abstract. Strong Ground Motions have a significant impact on structural analysis and design. Seismic evaluation is deemed necessary for the quality, reliability, and feasibility of existing and developing structures. Earthquakes are even more disastrous when a mainshock is succeeded and preceded by aftershocks of higher magnitude. Most of the time, these aftershocks are ignored during analysis. In the last few decades, steel buildings have been crucial to the construction sector. On the other hand, RCC frames are commonly designed structures for commercial purposes and cost-effectiveness. IS 1893 (Part:1)- 2002 Criteria for Earthquake Resistant Design of Structures and IS 13920- 2016 Ductile Detailing of RC Structures Subjected to Seismic Reactions are majorly used for Seismic Analysis. Steel bracings in the structuralsystem increase the ductility and stiffness of the frame structure. That can be arranged in a variety of ways, such as X, diagonally, alternatively, V, inverted V, K, etc. Cross bracings are used in this paper for the design of a typical multi-story (G+9) semi-rigid steel frame and a typical RCC frame is designed of the samedimensions. Both the frames are modelled and compared using ETABS software and a static nonlinear Time History analysis is executed to examine theperformance of both the frames under the Mainshock and Aftershock of Chamoli earthquake. The Mainshock and Aftershock results of the Chamoli earthquakeare extracted from the Centre for Engineering Strong Motion Research Ground Motion Database. Base shear, joint displacement, kinetic energy, and storydisplacement are a few of the variables of Time History Analysis that affect how well a building performs during mainshock and aftershock earthquakes. When analysing the outcomes, it is important to take into account that each of these factors significantly affects how a building responds to seismic loads.

Keywords: Time history analysis · Cross bracing system · RCC Building frame · Time history · ETABS software · Seismic response · Semi-Rigid Steel Frame · Chamoli Earthquake

1 Introduction

Man-made structures in seismic areas are exposed to a tectonic pattern of aftershocks, the mainshock, and aftershocks, as well as a single earthquake sequence. The mainshock causes aftershock events because of modifications in both static and dynamic stresses that actually occur during the seismic process. The same geographical area as a previous significant earthquake experiences an aftershock, a minor seismic event [1]. The mainshock that first occurred is categorized as a foreshock when an aftershock is greater than the mainshock and is given the function of the mainshock. Top displacements and story drift were shown to be considerably impacted by the mainshock aftershock ground motion sequence [2]. The study was carried out by ETABS. Due to their distinct seismic-tectonic background, only few areas on Earth are subject to several earthquakes in a fairly short period of time. It is still in its infancy to be able to design buildings to endure several tremors with significant ground shaking. There are currently no parametric systems that can forecast the consequences of many major earthquakes on structures. [3][4].

On Mar 29, 1999, at 00:35:13.4 hrs IST, a moderate earthquake with a magnitude of 6.8 rocked the Garhwal - Kumaun area of the Western Himalayas. The epicentre wassituated at 30.408N, 79.416E, and the focal length was assessed to be 21 kilometers. In 1991, a massive earthquake with a focal depth of 1kilometresrs had an epicentre at Uttarkashi at 30.74N and 78.79E. There are not many people living in the earthquake-damaged area, and the topography is extremely unlevel [11]. The most common building type in the region is stripped stone masonry in mud mortar with hefty yet elastic stone slab roofs. The earthquake generated difficult terrain with hills andlandslides, which obstructed numerous travel routes and hindered rescue and relief efforts [5][6].

The earthquake shook the surrounding region of Rudraprayag, the towns of Chamoli and Gopeshwar, and both. Most of the hardest-hit areas are located in the valley of theAlakananda River. The peak intensity on the damage-based MSK scale was observed around positions VII and VIII. The intensity of the shaking was quite unevenlydistributed. According to Indian rules for setting specifications for earthquake-resistant constructions, the impacted region is located in seismic zone V, which bears the highest level of risk. The system forecasts that earthquake zone V will have a magnitude of IX or higher. [7].

In this study, we compare the seismic responses of a G+9 multi-story RCC frame building (Model 1) with a semi-rigid frame with bracings (Model 2) during the Mainshock and Aftershock earthquakes separately. We also evaluate seismic metrics for the suggested structural models, such as story deflection, story drift, story shears, and story stiffness. We will use time history analysis with statistics from the Centrefor Engineering Strong Motion Research Ground Motion Database on Mar 29, 1999, GOPESHWAR STATION, CHAMOLI EARTHQUAKE (Latitude & Longitude:
30'24"N - 79'20"E) to examine the seismic features of the structure. [8]

2 Methodology

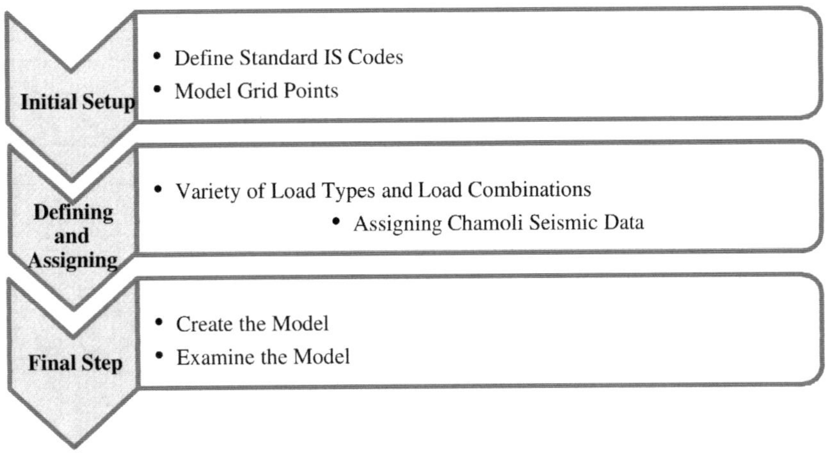

3 Building Parameters

A G+9 building is represented in ETABS v16 with a storey height of 3.1 m, a structural length of 20 m in one axis and 15 m in the other, and component sizes that vary depending on the design needs. For Model 1 and Model 2, the RCC slab has a thickness of 150 centimetres and a deck size of 3 inches, respectively. Models 1 and 2 were evaluated in accordance with IS 456-2000 and IS 800-2007, respectively. Seismic criteria are based on IS 1893-2016 [9][10].

3.1 Material Properties for Model 1

Table 1. Building Properties

Building Parameters	Dimensions
Plan dimension	20X15 m^2
No. of stories	9
Floor-to-floor height	3100 mm
Concrete	M30
HYSD	500
Columns	450*450
Column 2	450*500
Beam 1	230*400
Beam 2	230*450
Slab Thickness	150 mm

3.2 Material Properties for Model 2

Table 2. Building Properties

Building Parameters	Dimensions
Plan dimension	20X15 m^2
No. of stories	9
Floor-to-floor height	3100 mm
Beam size	ISWB 400 as per IS:800-2007
Column size	Two ISMC 400 (Backto Back) as per IS:800-2007
Bracing	ISA 150*150*15 as per IS:800-2007
Grade of Steel	Fe345
Deck Size	3 inches

3.3 Loading Conditions

For dead loads, we have considered IS 875 part 1 and for live loads, IS 875 part 2 accordingly and seismic analysis is carried out in accordance with the2016 edition of IS 1893 part 1.

Table 3. Load Calculation

Load	Data
Live load	2 kN/m^2 (IS 875-Part 2)
Super Dead load	1 kN/m^2 (IS 875-Part 1)
Floor Finish for Model 1	1.1 kN/m^2 (IS 875-Part 1)
Seismic zone	IV
Zone factor	0.24
Cladding Load for Model 2	5.0 kN/m^2 (IS 875-Part 2)
Acceleration due to gravity	9.81 m/sec^2
Response reduction factor	5
Importance factor	1.5
Scale factor (I.g/R)	1.962

IOP Conf. Series: Materials Science and Engineering 1273 (2023) 012011 doi:10.1088/1757-899X/1273/1/012011

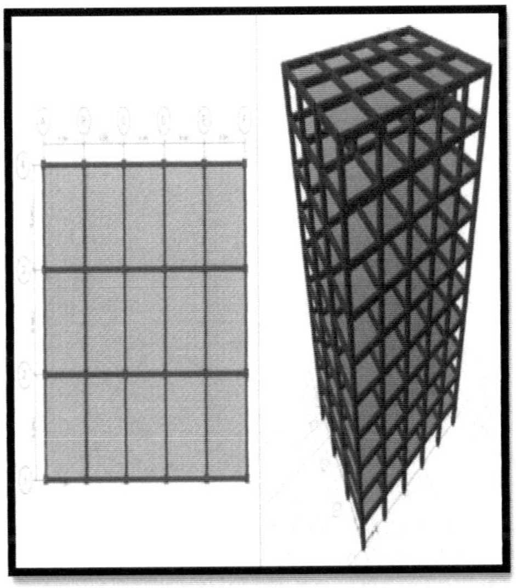

Figure 1. Plan and 3D shots for Model 1

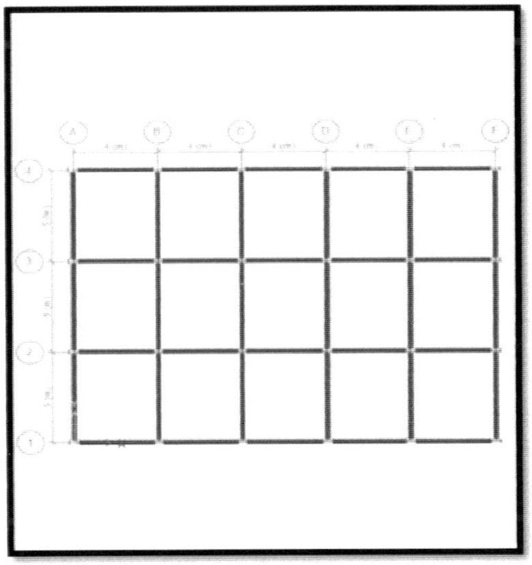

Figure 2. Plan for Model 2

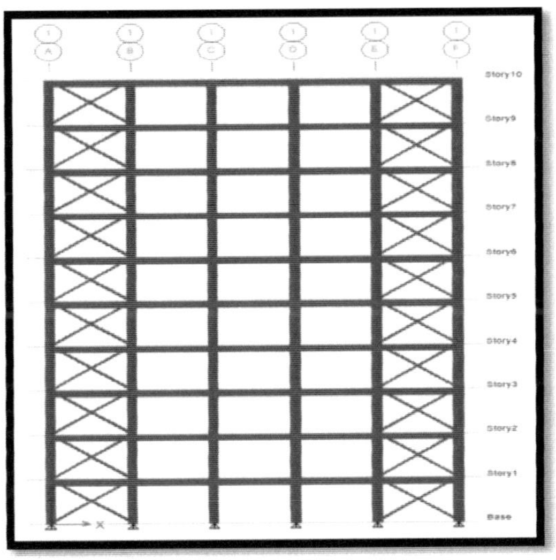

Figure 3. Elevation view for Model 2 from 'Section A'

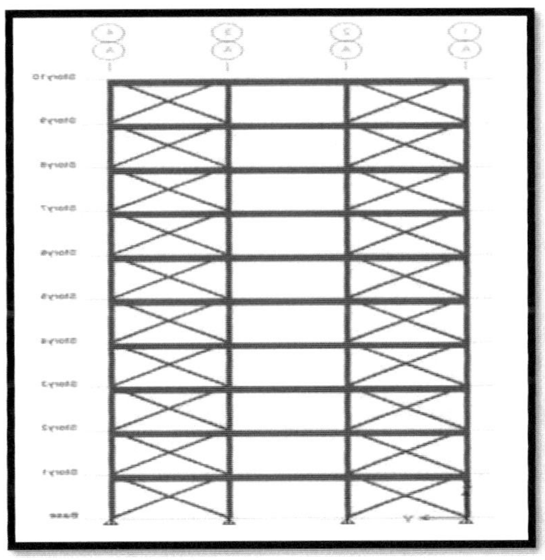

Figure 4. Elevation view for Model 2 from 'Section 1'

IOP Conf. Series: Materials Science and Engineering 1273 (2023) 012011 doi:10.1088/1757-899X/1273/1/012011

4 Mainshock Earthquake

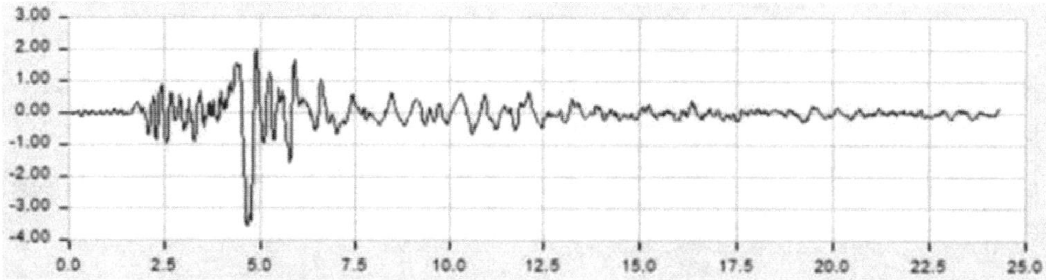

Figure 5. Data along the horizontal N-E direction showing Acceleration vs. Time

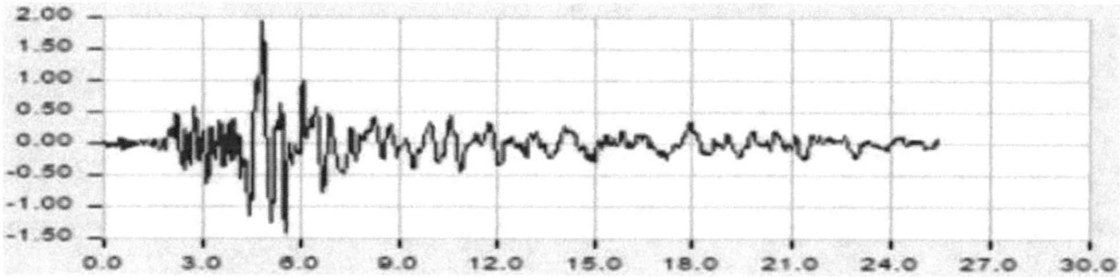

Figure 6. Data along the horizontal N-W direction showing Acceleration vs. Time

5 Aftershock Earthquake

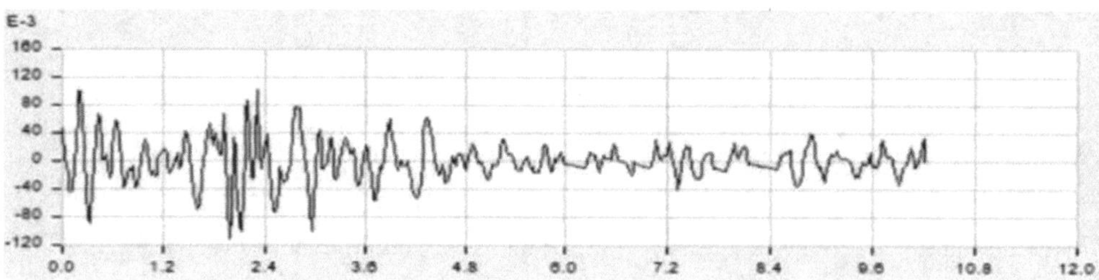

Figure 7. Data along the horizontal N-E direction showing Acceleration vs. Time

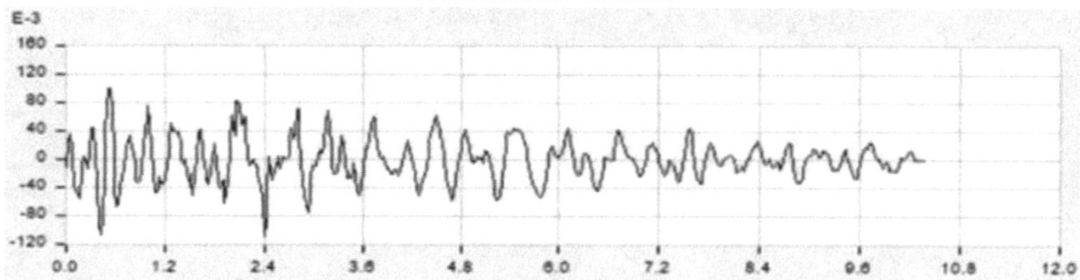

Figure 8. Data along the horizontal N-W direction showing Acceleration vs. Time

6 Results and Discussion

6.1 *Model 1*

6.1.1 *Story Displacement*

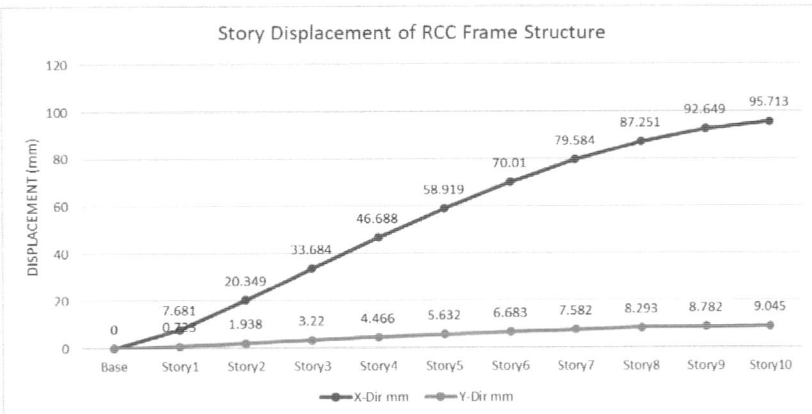

Figure 9. Storey Displacement for RCC frame structure

6.1.2 *Base Shear*

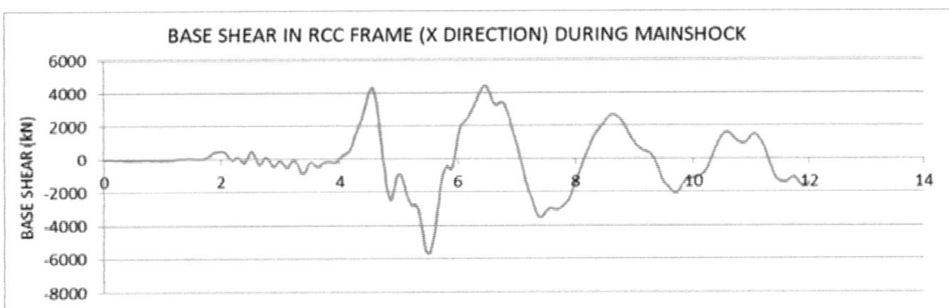

Figure 10. Base Shear during the Mainshock (X direction)

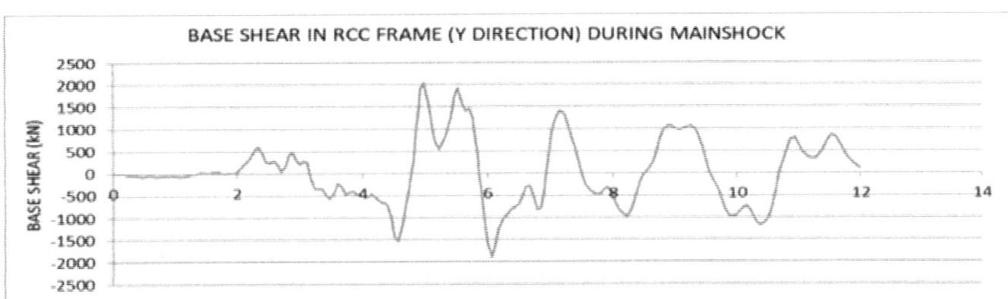

Figure 11. Base Shear during the Mainshock (Y direction)

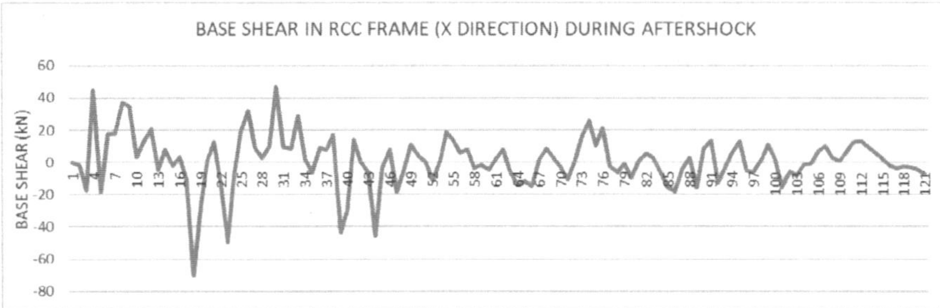

Figure 12. Base Shear during the Aftershock (X direction)

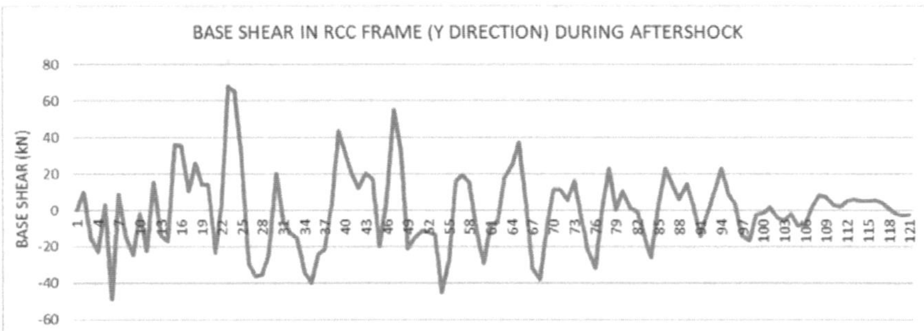

Figure 13. Base Shear during the Aftershock (Y direction)

6.1.3 Joint Displacement for Label 1 of Story 10

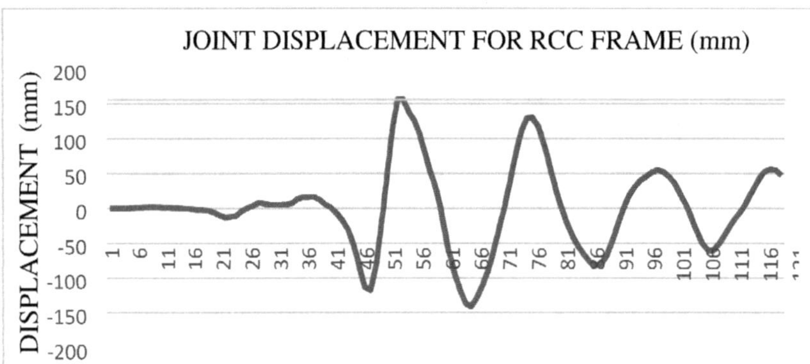

Figure 14. Joint Displacement during the Mainshock

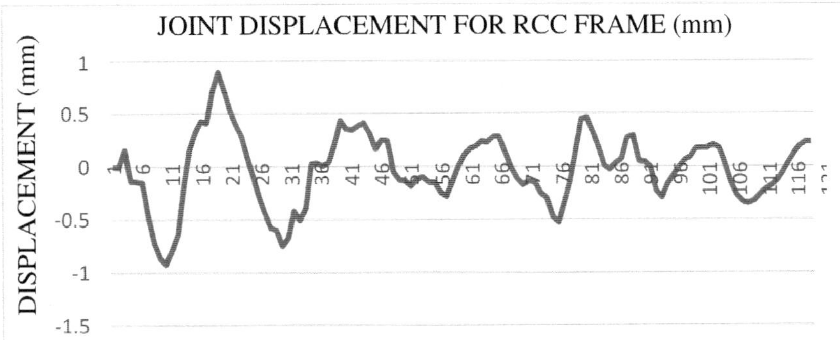

Figure 15. Joint Displacement during the Aftershock

6.1.4 Kinetic Energy Stored during Earthquake

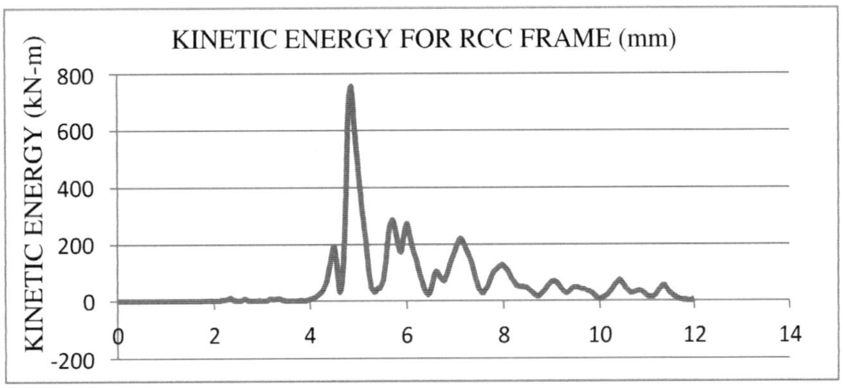

Figure 16. Kinetic Energy stored during the Mainshock

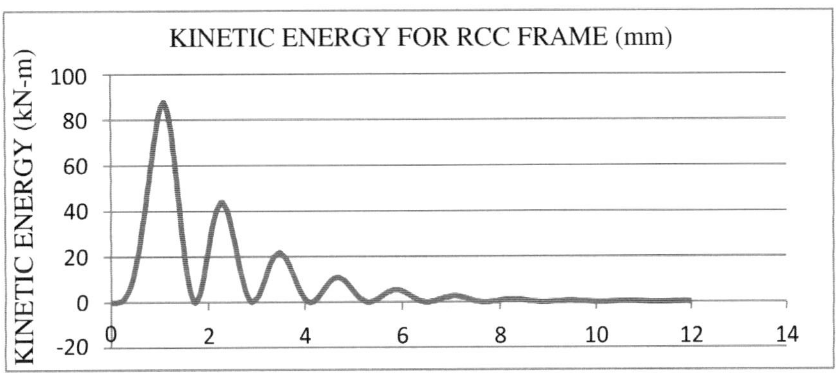

Figure 17. Kinetic Energy stored during the Aftershock

6.2 Model 2

6.2.1 Story Displacement

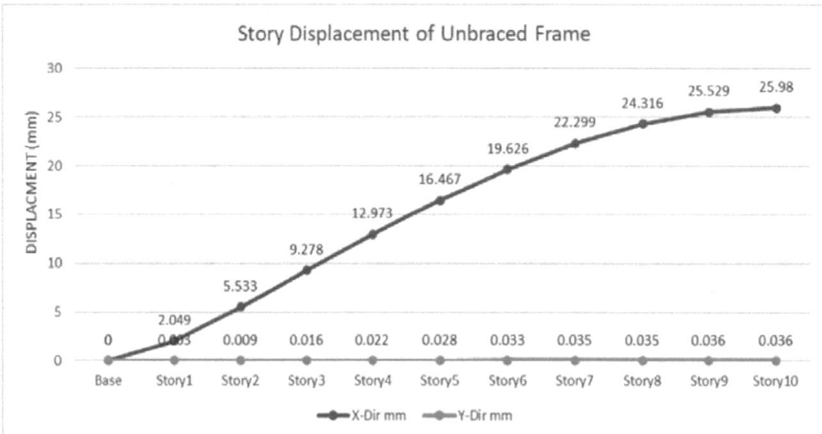

Figure 18. Storey Displacement for Unbraced semi rigid steel frame structure

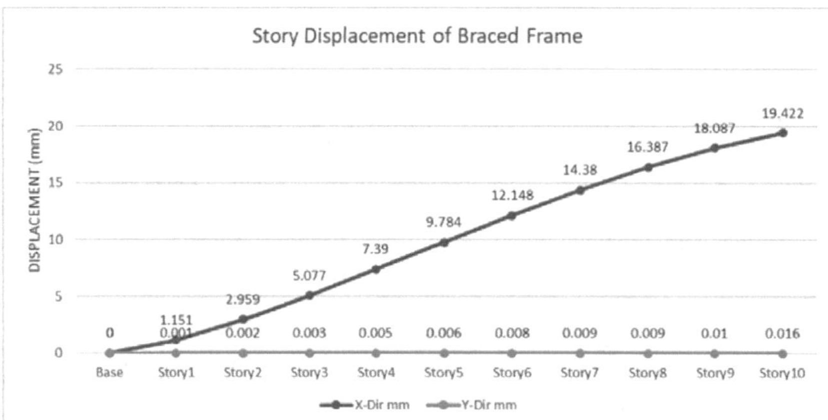

Figure 19. Storey Displacement for Braced semi rigid steel frame structure

6.2.2 Base Shear

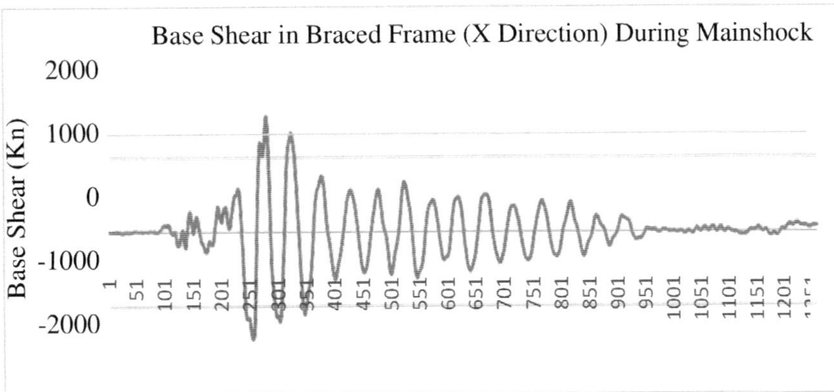

Figure 20. Base Shear in Braced Frame Mainshock (X Direction)

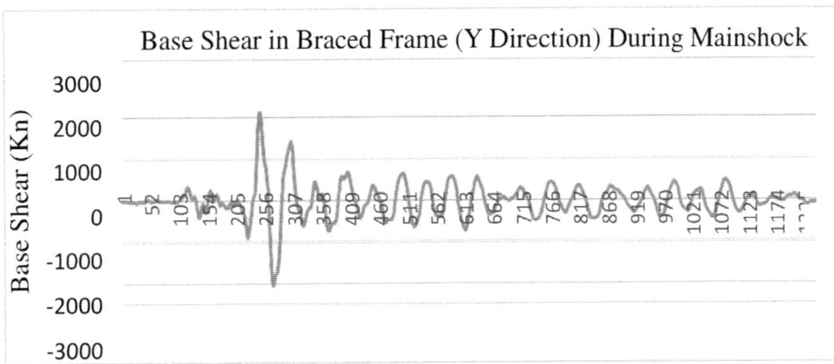

Figure 21. Base Shear in Braced Frame Mainshock (Y Direction)

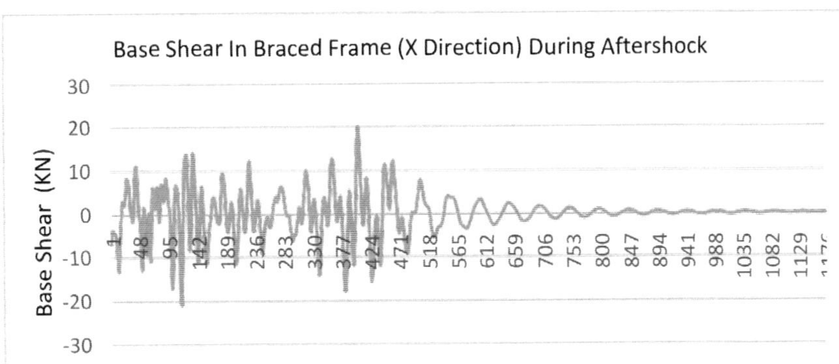

Figure 22. Base Shear in Braced Frame Aftershock (X Direction)

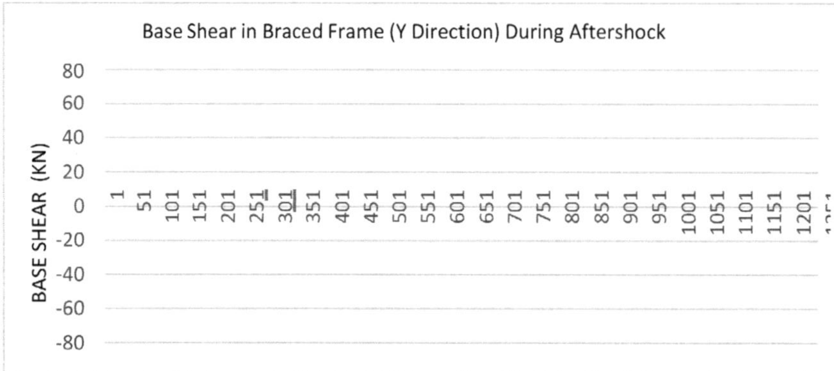

Figure 23. Base Shear in Braced Frame Aftershock (Y Direction)

6.2.3 *Joint Displacement for Label 1 of Story 10*

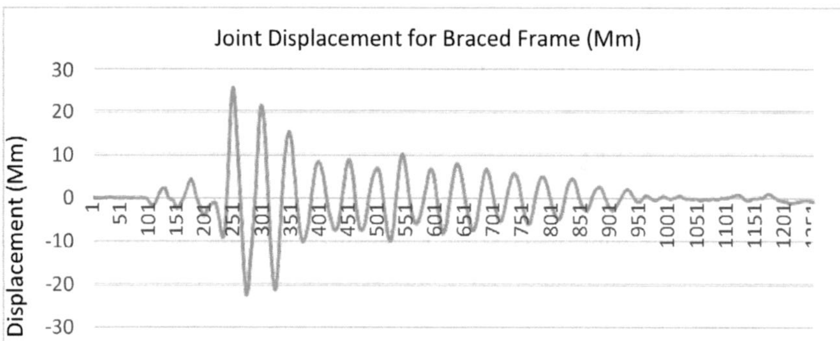

Figure 24. Joint Displacement during the Mainshock

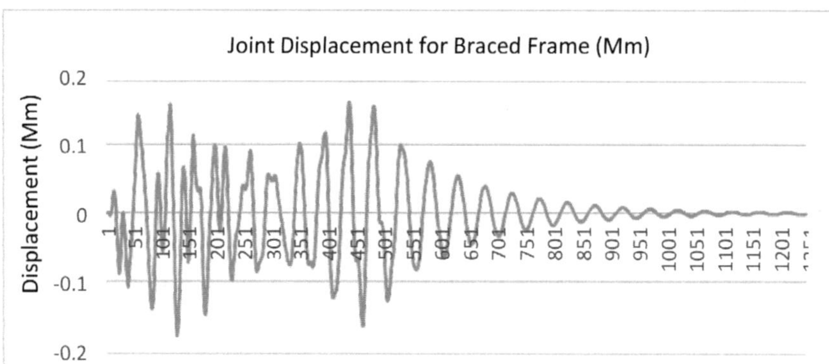

Figure 25. Joint Displacement during the Aftershock

6.2.4 Kinetic Energy Stored during Earthquake

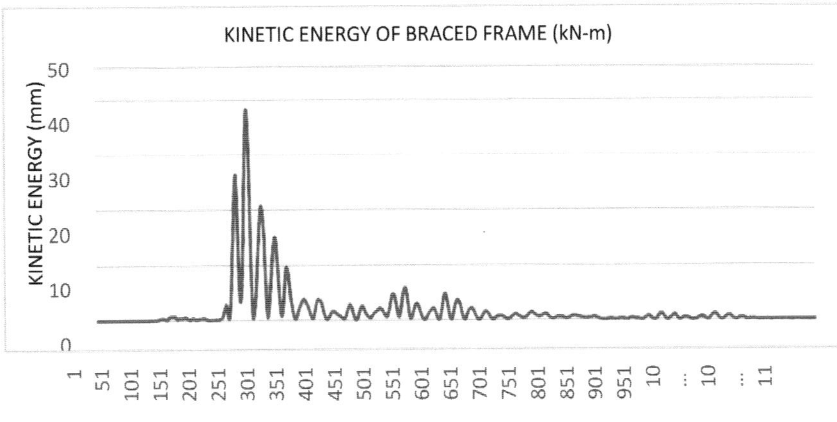

Figure 26. Kinetic Energy stored during the Mainshock

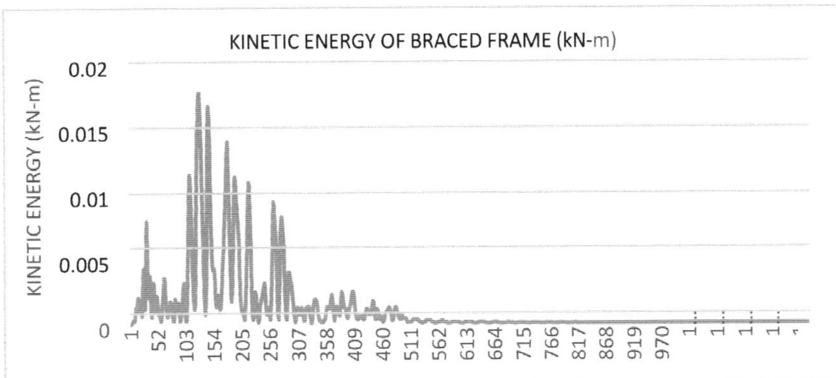

Figure 27. Kinetic Energy stored during the Aftershock

The following has been seen when comparing the Main Shock and Aftershock Earthquake on the G+9 RC Structure:

- According to IS-1893 (Part-1) of 2002, CI. 7.11.3, the storey drift in any storey due to the minimum specified design horizontal force, with partial loadfactor of 1.0, shall not exceed 0.004 times the storey height. Based on the obtained graphs, the highest drift is 0.019418 for Model 1 and 0.000772 for Model 2, both of which are well within the permitted limits [12].
- Maximum value of Story Displacement for Unbraced Semi Rigid Steel Frame is approx. 70 percent less than that for RCC Frame value. Moreover, Maximum Story Displacement for Braced Semi Rigid Steel Frame is approx. 80 percent less than RCC frame maximum story displacement values.

- Frequency of Joint Displacement and Base Shear is greater in case of Aftershock as compared to Main shock for both models.
- Maximum value of Joint Displacement of Storey 10 joint during Aftershockis approx. 2 percent of joint displacement during Main shock for RCC frame and approx. 0.65 percent for unbraced semi rigid steel frame and its approximately negligible for Braced semi rigid steel frame.
- Maximum Kinetic Energy of the building during aftershock is approx. 11.52 per-cent of Kinetic Energy stored during Main shock for RCC frame and its 9.4 per- cent for unbraced frame 5.5 percent for braced frame.
- Maximum Base Shear during aftershock for RCC frame is approx. 1.5 percentofbase shear during Main shock for RCC frame and approx. 0.22 percent for unbraced semi rigid steel frame and it's approximately negligible for Braced semi rigid steel frame.
- Storey drifts, storey displacement and stiffness came out to be in the permissible limits as recommended by the IS 1893 code.

7 Conclusion

Conclusions for the structural framework of a G+9 construction in India with variouskinds of soil in earthquake zone IV have been established based on time history study.

- According to the discussed results, aftershocks are unavoidable earthquake sequences that should always be considered when building earthquake-resistant structures, especially when the structure has previously sustained damage from afar more powerful main shock [13].
- The construction's initial level is where the story shear begins low and rises to the top storey. The size of the story shear increases as a building's height climbs.
- To prevent the damage due to mainshock and aftershock sequence, the base isolation system, which isolates the structure (superstructure) from the base, can be provided (foundation or substructure). By separating the buildingfrom its base, the amount of energy transmitted towards the building during aseismic event is significantly reduced.
- Aftershocks with a higher frequency and lower Richter scale strength can have catastrophic repercussions for any building that has already undergone aprimary shock.
- Braced Frame Semi Rigid Steel Structure showed best performance in every aspect followed by unbraced semi rigid steel frame structure. RCC frame per-formed poorly due to less ductility and light weight [14].
- The construction's initial level is where the story shear begins low and rises to the top storey. The size of the story shear increases as a building's height climbs.

IOP Conf. Series: Materials Science and Engineering 1273 (2023) 012011 doi:10.1088/1757-899X/1273/1/012011

- Time history analysis, a complicated technique, may be used to analyse how well a structure and its constituent pieces, such as the vertical supports,beams, and slab, are performing. The structural capacity of the structure may be assessed by selecting appropriately selected ground vibration signals from a previous earthquake.
- Structures can be kept stable by using cross bracing during seismic events like earthquakes and windy conditions.

8 Future Scope

- Torsion effects, joint reaction, joint displacement and velocity may be analyzed for further discussions.
- Comparative study of different types of the bracing system at different location inthe semi rigid steel frame structure.

- **References**

[1] A. S. Patil and P. D. Kumbhar, "Time History Analysis of Multistoried Rcc Buildings for Different Seismic Intensities," Int. J. Struct. Civ. Eng. Res., vol. 2, no. 3, pp. 195–201, 2013.

[2] G. B. Katti and B. S. Balapgol, "Seismic Analysis of Multistoried RCC Buildings Due to Mass Irregularity by Time History Analysis," vol. 3, no. 7, pp. 614–617, 2014.

[3] B. Lanjewar and P. A. Khedikar, "ISSN NO : 0022-1945 'Seismic Response Of RC Build-ing Under Main shock-Aftershock Sequence' Page No : 1409 ISSN NO : 0022-1945 Page No : 1410," vol. XII, no. 1409, pp. 1409–1419, 2020.

[4] S. Kumitani and T. Takada, "Probabilistic assessment of buildings damage considering aftershocks of earthquakes," J. Struct. Constr. Eng., vol. 74, no. 637, pp. 459–465, 2009, doi: 10.3130/aijs.74.459.

[5] A. Massumi, K. Sadeghi, and H. Ghaedi, "The effects of mainshock-aftershock in successive earthquakes on the response of RC moment-resisting frames considering the influenceof the vertical seismic component," Ain Shams Eng. J., vol. 12, no. 1, pp. 393– 405, Mar. 2021, doi: 10.1016/j.asej.2020.04.005.

[6] W. Huang, J. Qian, and Q. S. Fu, "Damage Assessment of RC Frame Structures under Mainshock-Aftershock Seismic Sequences."

[7] J. Ruiz-García and J. C. Negrete-Manriquez, "Evaluation of drift demands in existing steel frames under as-recorded far-field and near-fault mainshock-aftershock seismic sequences," Eng. Struct., vol. 33, no. 2, pp. 621–634, Feb. 2011, doi: 10.1016/j.engstruct.2010.11.021.

[8] J. Ruiz-García, "Issues on the Response of Existing Buildings Under Mainshock-After- shock Seismic Sequences."

[9] A. Ilyas, M. A. Azeem, and H. Mohiuddin, "seismic response of reinforced concrete buildings under mainshock – Aftershock earthquake sequence," Int. J. Civ. Eng. Technol.,vol. 9, no. 4, pp. 647–659, 2018.

[10] S. D. V. M. V Waghmare, "Nonlinear Analysis of RC Structure under Multiple Earth- quakes," vol. 5, no. September, 2019.

[11] Shesalu D.Vadeo1, M.V.Waghmare ,"Nonlinear Analysis of RC Structure under Multiple

Earthquakes", International Journal for Modern Trends in Science and Technology, Volume: 05, No: 09, (2019), 2455-3778
[12] Mya Mya Aye, P.Narasimharao, "Seismic Analysis of Multistoried RCC Building due to Mass Irregularity by Time History Analysis", Journal of Emerging Technologies and Innovative Research, Volume 4, No.12, (2017),2349-5162
[13] A.Krishna Srinivas, B.Suresh, A.Madhusudhan Reddy, "Time History Analysis Of Irregular Rcc Building For Different Seismic Intensities" International Journal of Scientific & Engineering Research Volume 8, Issue 7, (2017), 736, 2229-5518
[14] Prakriti Chandrakar, Dr. P.S.Bokare, "Earthquake Analysis of G+10 Building using Re- sponse Spectrum Method and Time History Method - A Comparison, International Journalof Science and Research (IJSR),(2015), 2319-7064
[15] Kaushal Vijay Rathod, Sumit Gupta, "A Nonlinear Time History Analysis Of Ten Storey Rcc Building" International Research Journal of Engineering and Technology (IRJET) Vol. 07, No. 06, (2020), 2395-0056
[16] Abdelnaby, A. E. [2016] "Fragility curves for RC frames subjected to Tohoku mainshock aftershocks sequences," Journal of Earthquake Engineering, (2016)11. R. Narasimhan and K. eMaging Solutions, "Estimation of source parameters of Chamoli Earthquake, India," 2001.

Estimation of Energy Absorption Capacity and Geotechnical Characterization of Silty-clay Soil Blended with Tire Fibres

Shiv Shankar Kumar[1] and Mani Shankar[2]
[1,2]National Institute of Technology, Patna, Bihar – 800005, India
[1]k.shiv.ce@nitp.ac.in,[2]manis.pg20.ce@nitp.ac.in

Abstract. This article presents the compaction, consolidation, unconfined and confined characteristics of tire-fibre blended silty-clay at different investigating parameters. The compaction tests were performed to evaluate the Maximum Dry Density (MDD) and Optimum Moisture Content (OMC) of soil-tire fibre mixture at different percentage of tire fibre (0%, 10%, 20%, 30% and 40% by weight). It was found that with increasing the amount of tire-fibre MDD of mixture decreased significantly whereas, OMC increases marginally. Both, unconfined compression strength (UCS) and confined triaxial tests were performed on the soil-tire mixes specimen prepared at MDD and OMC. The result shows that with increasing percentage of tire fibre to the silty-clay soil, the ductility of the soil-tire mixture increases whereas; the maximum shear stress at the failure is reduced. Based on UCS tests, the Energy Absorption Capacity (EAC) of soil-tire mixture is increased increased by 55% at 10% of tire fibre, for strain rate of 1.25mm/min, whereas; on further increase in TF from 10% to 30%, EAC decreased by 44%, 32%, 52% at the shearing rate of 0.62mm/min, 1.25mm/min and 2.5mm/min, respectively. However, based on the confined triaxial test results, the angle of internal friction of soil-tire mixture decreases with increasing tire contents; whereas, cohesion decreased up to 10% of tire fibre and beyond that it increases. EAC was found to be increased up to 20% of the tire fibre and further addition shows reduction in energy absorption capacity.

Keywords: Silty-clay, Tire fibre, Energy absorption capacity, Oedometer, UCS, Confined compression test

1 Introduction

Due to the rapid growth of economy as well as the transport facility, the number of discarded tires has also been increased significantly and it has also been estimated that the discarded tires might grow up to 1.2 billion per year by 2030. Therefore, it is quite important to reuse of these waste tires in order to decrease the environmental as well as health impact on living things since, the tire contains different chemical compound such as elastomers, polyisoprene, polybutadiene, and styrene-butadiene (Trouzine et al.2012). Al-Tabbaa et al. (1997) have performed the compaction test on clay (kaolin)-tire mixes and repotted that the Maximum Dry Density (MDD) decreased with by increasing shredded tire however, the Optimum Moisture Content (OMC) has negligible effect on the addition of shredded tire. It was also observed that the hydraulic conductivity of clay (kaolin)-tire mixes reduced significantly by adding the shredded tire. Further, Al-Tabbaa and Aravinthan (1998) estimated the hydraulic conductivity of clay blended with paraffin treated rubber tire of size 4-8 mm utilizing distilled water, acidic water and paraffin and, it was reported that the hydraulic conductivity of clay reduced significantly (nearly 50%) when 15% tires is added to clay with paraffin. Cokca and Yilmaz (2004) have measured the compressibility characteristic of bentonite added fly ash and rubber fiber mixes using Oedometer tests. It was reported that the initial void ratio and swelling pressure decreased with increase of rubber fiber content to the bentonite added fly ash. It was also reported that the hydraulic conductivity of fly ash increased by increasing percentage of rubber tire whereas, decreased by adding bentonite content. Further, it was also reported that the Unconfined Compression Strength (UCS) of bentonite added fly ash and rubber fiber mixes was found to be decreased with increase in rubber fiber. Ozkul and Baykal (2006) have observed the strength and deformation characteristics of kaolinite clay and tire buffing mixes under drained and undrained conditions. It was found that the failure strain of reinforced specimen was much higher than that of unreinforced specimens whereas, the strength of reinforced specimen was slightly lesser than that of the unreinforced specimen. It was also observed that the drained shear strength of the clay was unchanged with increasing rubber buffing. Mokhtar and Chan (2007) have performed UCS tests on clayey soil mixed with cement and tire chips of 2 mm size. It was reported that the addition of tire chips increases the strength of cemented clay in comparison to the strength of without cemented clay. Marefat and Soltani-Jigheh (2011) have performed the consolidated undrained triaxial tests to investigate the behavior of clay-tire chips mixes. It was reported that the shear strength of clay doesn't show reduction in the shear strength by adding tire chips more than 20-30%. It was also observed that the secant modulus decreased whereas, excess pore water pressure of clay soil increased up to 10-20% of tire chips and further decreased, with the increase of tire-chips contents. Sarvade and Shet (2012) have investigated the influence of crumb rubber (smaller than 1.18 mm size) on clay soil added cement and lime. It was reported that the liquid limit of clayey soil decreases continuously with the increase of crumb rubber contents. Similar responses were observed for the plastic limit and shrinkage limit. Asadzadeh and Ersizad (2013) have examined the effect of tire chips on the consolidation characteristics of low-plastic clay and, it was reported that the swelling index increased whereas, compression index decreased by increasing the tire-chips content to the clay soil. Further, an increase in cohesion and decrease in internal friction angle was also reported when percentage tire chips increased. Soltani-Jigheh et al., (2013) were carried out undrained triaxial tests on the compacted clay-tire mixtures to examine its behaviors and, it was reported that the shear strength decreases by increasing tire content. It was also reported that the adding tire-chips to the clay leads to increase deformability of mixtures considerably. Prasad et al. (2014) have studied the influence of crumb rubber (425–600 μm) on the compaction behaviour of two types of clayey soils and reported that MDD and OMC decreased for both types of clay with the addition of rubber fiber. Daud et al. (2015) have examined the influence of shredded tires (size of 1–5 mm) on Atterberg limit of peat and clayey soil and, it was found that there is an increase in liquid limit and plastic limit of clayey soil by the inclusion of shredded tires up to 20% however, for peat soil, the liquid limit and plastic limit shows decreasing trend with the addition of shredded tires. Further, the compressibility of the clayey soil mixed with crumb rubber has been studied by Signes et al. (2016) and reported that the swelling index increases with the addition of crumb rubber whereas, the compression index remained unchanged.

The hydraulic conductivity of clayey soils with the addition of crumb rubber (in the order of 5%, 10%, 15%, and 20%) have been studied by Ravichandran et al. (2016) and reported that the hydraulic conductivity of the bentonite clay soil is increased with the increase of crumb rubber fraction, which is attributed to the increase of flow channel in between crumb rubber-bentonite mixture. Yadav and Tiwari (2017) have reported that the swelling pressure as well as UCS of clay decreases with increasing the percentage of rubber fiber. Yadav and Tiwari (2018) studied the compaction and strength characteristics of clayey soil mixed with crumb rubber and cement for geotechnical engineering applications and, it was reported that MDD and OMC were decreased with the inclusion of tire fiber whereas, UCS of soil-tire mixture has been increased up to 5% of tire fiber and further addition of tire fiber resulting decrease in the UCS value. It was also observed that MDD, UCS and the unsoaked CBR value decreased with the increase of crumb rubber (Saberian et al. 2019). Mukherjee and Mishra (2020) were investigated the effect of tire fiber on the shrinkage, hydraulic conductivity and shear strength of sand-bentonite (SB) mixture. It was reported that the hydraulic conductivity of the mixture increased and volumetric shrinkage (VS) decreased by increasing the tire fiber content. Triaxial test results suggested that shear strength parameters and Young's modulus enhanced with increasing percentage of tire fiber. Jastrzebska and Tokarz (2021) have performed the cyclic triaxial tests on consolidated undrained sample to examine the effect of rubber-tire waste on the strength characteristics of two different soils kaolin and swelling red clay. It was reported that the shear strength of rubber-tire mixed soil decreased due to cyclic loading. It was observed that the effects of shear strength depended on the mixture content and size of the rubber-tire waste.

Based on the aforementioned literature, it can be stated that the several researchers have studied the influence of tire (may be of different sizes) on the sand, clay and expansive soil. It was reported that MDD, hydraulic conductivity, swelling pressure, compressibility and shear strength of clayey soil affected significantly by adding tire fiber. During literature study, it was found that the results obtained by one researcher are contradicting to the others. It was reported that the hydraulic conductivity was increased with tire fiber however; some studies were also reported that hydraulic conductivity decreases with tire fiber. It was also reported that the inclusion of tire enhanced UCS of clay however; other study showed that UCS of expansive clay dropped with the addition of tire fiber. It was also noticed that the inclusion of waste tire fiber increased the shear strength, but on the same time studies also reports that the decrease in shear strength with waste tire fiber. Moreover, it was found the study on Silty-clay added with tire fiber was very limited. Therefore, this study focus on the behaviour of silty clay mixed with tire fibre at different investigating parameters for geotechncail engineering.

2 Materials used

The materials utilised in this investigation were classified as silty clay (i.e., cohesive silty clay (MC)), tire fibre (TF), and tire fibre-soil (CTF) mixtures.

2.1 Silty Clay Soil

Silty-clay soil, shown in Fig. 1a comprises of fine silt and clay particles was collected from Madhubani district, Bihar. Before performing the different test, the index properties such as: particle size distribution, Atterberg's limit, maximum dry density, optimum moisture content, specific gravity, moisture content, etc. have been evaluated which helps to access the physical and engineering behaviour of a soil. To find out the different sizes of the soil particles within the soil mass, hydrometer analysis was carried out since soil was found to be in the range of fine soil category. The particle size distribution, obtained from hydrometer analysis, depicted in Fig. 1c is a graphical representation of sieve analysis that plots the percentage finer against particle size on a logarithmic scale for the soil and tire fibre used in the present study. Based on the grain size analysis, as shown in Fig. 1c, it was observed that the soil contains nearly 20% clay fraction (passing through 0.002 mm) and 60% silt. This characterization of soils are utmost important since each and every soil respond differently under different loading and testing conditions. The specific gravity (G_s) of soil was found to be 2.57. Further, the consistency limits were determined for the soil sample. This limits are mainly used for clayey soils and is related to water content i.e., how with the change in water content the consistency of soil changes. According to Atterberg, four different consistency stages exist within the soil mass such as: solid stage, semi-solid stage, plastic stage and liquid stage. The liquid limit (w_L), plastic limit (w_P) and Shrinkage limit (w_s) was found to be 38.87%, 22.82% and 21.59%, respectively. Plasticity Index (PI) of soil found to be 16.05%, which by indicates the medium plastic soil and cohesive in nature. This result is also consistent with the range of PI of silty-clay reported by Bowles (1996) as mentioned in Table 1.

Table 1 Classification of soil based on Plasticity Index

Plasticity index (%)	Soil type	Degree of plasticity	Degree of cohesiveness
0	Sand	Non plastic	Non cohesive
< 7	Silt	Low plastic	Partly cohesive
7-17	Silt clay	Medium plastic	cohesive
> 17	clay	High plastic	cohesive

2.2 Tire Fibre

Tire fibre used in the present study, shown in Fig. 1b, was collected from a tire processing unit established in Guwahati, Assam. Based on the sieve analysis, it was observed that the tire fibre comprises different size ranging from 0.075 mm to 2.36 mm in various proportions with most of size laying in the range of 0.6 mm to 1.18mm. The specific gravity of the fibres was determined using pycnometer, and was found to be 0.92. According to the literature study, the range of specific gravity was in the order of 0.83 to 1.21, depending on the size of tire particles (Bekhiti et al., 2019; Reddy et al., 2017; Fu et al., 2014; Kim 2008; Rao and Dutta 2006; Fosse et al., 1996).

(a)

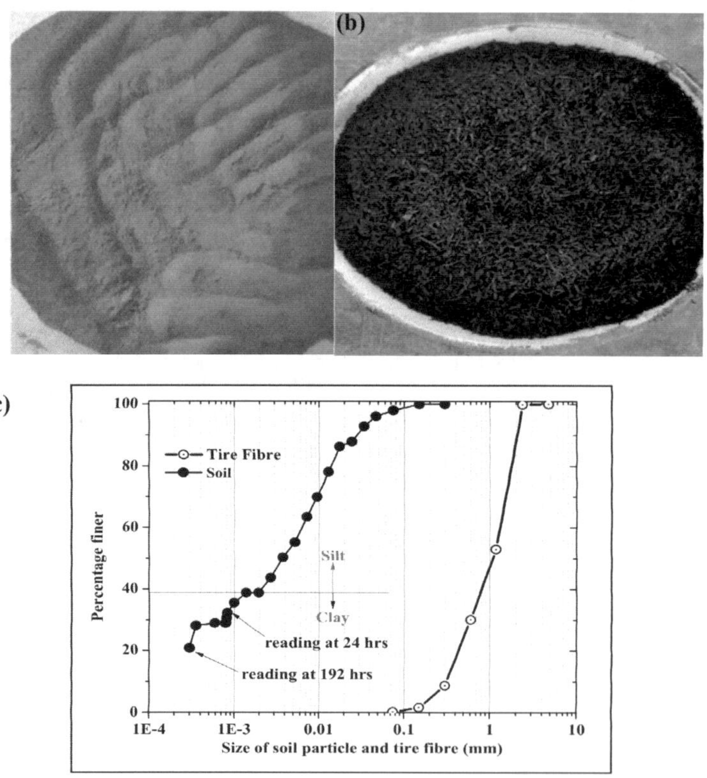

c)

Figure 1. Photograph of (a) silty-caly soil (b) tire fibre (c) particle size distribution for soil and tire fibre

doi:10.1088/1757-899X/1273/1/012012

2.3 Soil-Tire Fibre Mixes

For the present study the soil sample was mixed with tire fibres in different proportions ranging from 0% to 40% by weight. The specific gravity (G_s) of the mix was determined by calculating the weighted mean of G_s of soil and tire fibre. Liquid limit for each soil-tire mix were also evaluated to check the consistency with increase in tire content in the mix. It can be observed from Fig. 2a that the liquid limit of soil-tire mixes decrease as the tire content is increased. Furthermore, it is required to compact the soils to a dense state in order to get satisfactory engineering parameters for design and construction, as loose soil may not provide satisfactory results. Figure 2 b shows the compaction test plot at different tire fibre content. It can be observed from Fig. 2b that maximum dry density (MDD) was found to be 1.92 g/cc corresponding to optimum moisture content (OMC) = 13.4%, at 0% tire fibre. The obtained results are in accordance with finding by Cetin et al (2006), Srivastava et al (2014) and Oikonomou and Mavridou (2009). This decrease in MDD can be attributed to the fact that tire have lower specific gravity than the soil. Furthermore, there is a slight decrease of 2% in MDD value as the tire content is increased from 30 % to 40% as compared to initial decrease of around 7% with each subsequent 10 % increase in tire content from 0% to 30%.

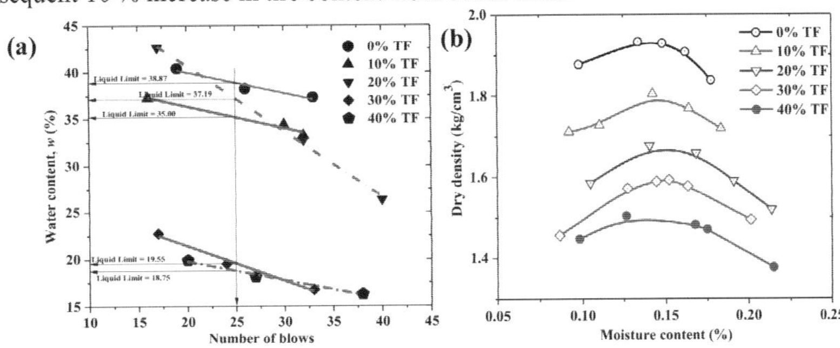

Figure 2. (a) Liquid limit and (b) Compaction curve of soil-tire mix, at different fibre percentage

3 Results and Discussion

This section discusses the results obtained from the tests like oedometer, unconfined compression test and confined compression test at different investigating parameters such as tire content (0%, 10%, 20%, 20%, 40%), shearing rates (0.62mm/min, 1.25mm/min, 2.5mm/min), confining pressure (50kPa,100kPa, 200kPa) to observe the stress strain response, compressibility characteristics, shear strength parameter and the energy absorption capacity of soil-tire mixes.

3.1 Effect on Compressibility Characteristics on Soil-tire Mix

Figure 3 presents the variation of void ratio (e) during the application of loading (σ) at different percentage of tire mix with soil. It can be clearly observed that with increase in tire content, the slope of compression curve increase indicating an increase in compression index (C_c). Moreover, there is a steep increase in the slope as the tire content is increased. One more interesting observation that can be made is that at higher tire content, the rate of increase in C_c value is more prominent.

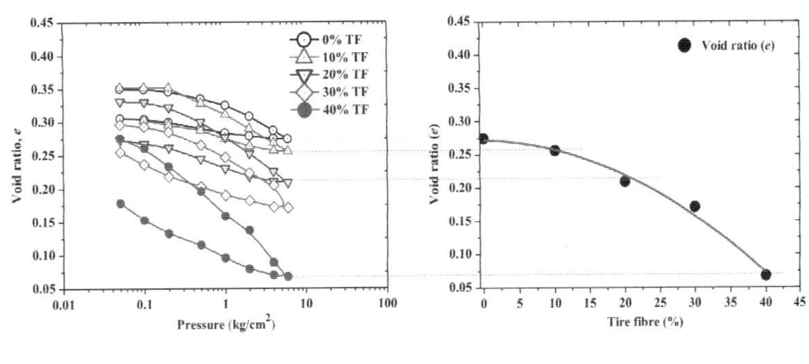

Figure 3. Loading unloading curve of soil tire mix from Oedometer test

3.2 Effect of Shearing Rate on the Behaviour of Soil and Soil-tire Mix

Figure 4 (a-e) presents stress- strain response of samples containing soil with tire fibre mixed in the proportion of 0-40% at different shearing rates: 0.62mm/min, 1.25mm/min, 2.5mm/min. Figure 4a shows response of stress-strain of soil only (0% tire fibre) prepared at MDD and OMC. It indicates that as the shearing rate increases the peak shear stress of soil reduces. The peak value of shear stress reduced by 16% when the strain rate increased from 0.62mm/min to 1.25mm/min whereas, the same has been reduced by 31% from 0.62mm/min to 2.5 mm/min. It can also be seen that the soil sample prepared at MDD and OMC reflect brittle failure behavior corresponding to low axial strain (less than 6%). Figure 4b presents the stress-strain response of soil-tire mixes (10% tire fibre). It can be seen that the effect of strain rate is not significant on the stress-strain response at 10% tire fibre mixed soil. It was also noticed that by adding the tire fibre the soil reflects the ductile behavior and achieving the axial strain value of 12.5% before its fail. Figure 4c presents the stress-strain response of 20% tire fibre mixed soil. Similar response, in terms of peak shear strength, is observed in Fig. 4c, but it differs in the axial strain at which the samples fail. The sample fails at around 9% of axial strain, which is 28 % lower than the axial strain achieved at Fig. 4b. Figure 4d presents the stress-strain response of 30% tire fibre mixed soil. The response of soil-tire mix shows the ductile failure. However, the peak shear stress was nearly 1.6 kg/cm^2, which is significantly lesser than the value observed in Fig. 4a, Fig. 4b and Fig. 4c.

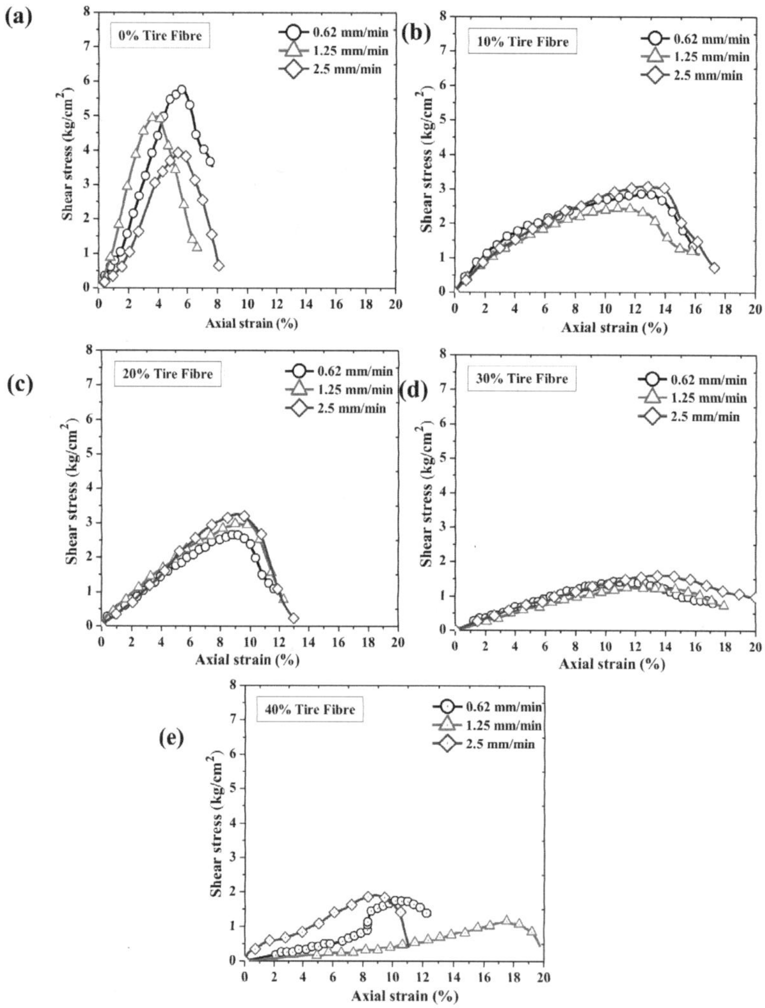

Figure 4. (a-e) Stress train response at 0 to 40% tire mix

Figure 4e presents the stress-strain response of 40% tire fibre mixed soil. From Fig. 4e, it can be seen that at extreme loading rates (0.62mm/min, 2.5 mm/min), the peak shear strength was nearly 1.8 kg/cm^2 and at the loading rate of 1.25 mm/min, the strength decreases by 11% and the axial strain at failure increases by around 98%.

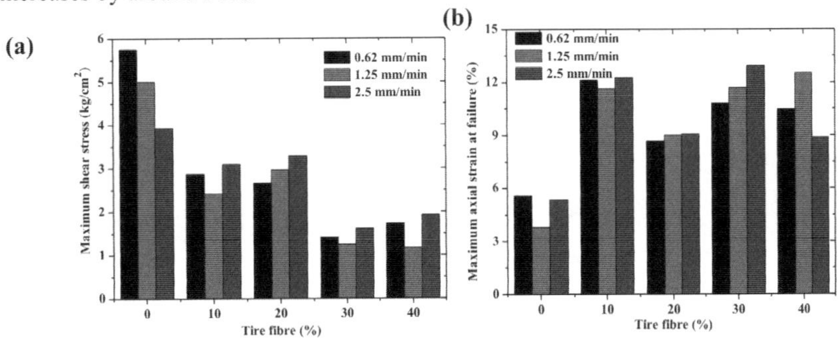

Figure 5. Variation of (a) maximum shear stress (b) maximum axial strain with tire fibre at different strain rate

Figure 5a indicates the variations of peak shear stress with tire fibre at different strain rate. From Fig 5a, it can be observed that the peak shear stress reduced significantly i.e., by 54% and 48% up to 10% TF mix for loading rate of 0.62mm/min and 1.25 mm/min, respectively. On further increase in TF to 20% there is a little increase of 17% and 4% in shear strength for 1.25mm/min and 2.5 mm/min loading rates. Further, on increasing TF by 30 %, the peak shear stress increases by 40%, 70% and 62% for three loading rates 0.62mm/min, 1.25 mm/min and 2.5 mm/min, respectively. Figure 5b presents the maximum axial strain achieved by the soil specimen at different percentage of tire before its fail. It can be observed that the soil sample without tire fibre indicates brittle failure whereas, with increase in the percentage of tire fibre, the ductility of soil specimen is increased. However, the ductile behaviour of soil is more prominent at 10% of tire fibre for all value of strain rate

3.3 *Effect of Confinement on the Behaviour of Soil and Soil-tire Mix*

Figure 6 presents the stress-strain response of soil and soil-tire fibre mixture at different confining stress (0.50kg/cm^2, 1kg/cm^2, 2kg/cm^2). The shearing rate of 1.25mm/min was employed for all the samples. Figure 6a shows stress strain response for 0% TF mix at different confining pressures obtained from Confined Compression Test (CCT). It can be observed that the maximum shear stress, at confining stress = 0.5 kg/cm^2, was found to be 8.5 kg/cm^2 and, nearly same value was observed for the maximum shear stress, at confining stress = 1.0 kg/cm^2. However, at confining stress of 2kg/cm^2, the maximum shear stress was found to be nearly 11 kg/cm^2. Thus, it can be stated that the maximum shear stress of soil (with 0% TF) is increased by 29% with increase in confining stress from 0.5 kg/cm^2 to 2.0 kg/cm^2. Figure 6b shows stress strain response for 10% TF mix at different confining pressures from CCT. It can be seen that with the increases in confining stress, shear stress also increases. It is noticed that the maximum shear stress is increased by 40% and 120%, when the confining stress increased from 0.5kg/cm^2 to 1kg/cm^2 and from 0.5kg/cm^2 to 2kg/cm^2, respectively. Figure 6c shows the stress-strain response for 20%TF at different confining pressures also indicating that with increasing confining stress the maximum shear stress also increases, which is similar to the response obtained in Fig. 6b. It is seen that the maximum shear stress is increased by 20% and 60%, when the confining stress increased from 0.5kg/cm^2 to 1kg/cm^2 and from 0.5kg/cm^2 to 2kg/cm^2, respectively. Figure 6d shows the insignificant effect of TF (40% TF) since all the specimen sheared at all three confining pressures reflects same value of maximum shear stress i.e., nearly 3kg/cm^2. Moreover, it can also be noticed that the ductility of soil increases with increasing tire fibre whereas, the maximum shear stress is reduced which results in the reduction of strength of soil.

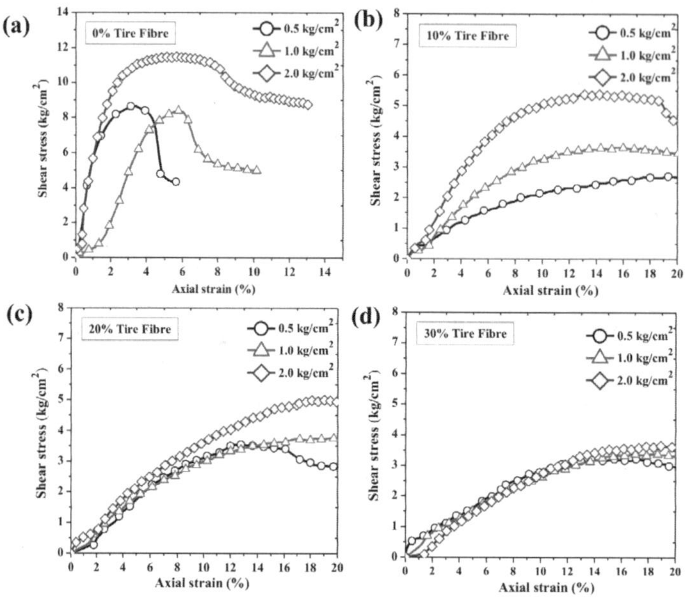

Figure 6. (a-d) Stress strain response of soil and soil-tire fibre mixes at different confining stress

Figure 7 (a, b) variations of peak shear stress and maximum axial strain achieved by soil mas before its fail at different confining stress for varying range of TF i.e., from 0%-30%. Figure 7a presents the bar chart between max shear stress and percentage tire fibre for different confining pressure.

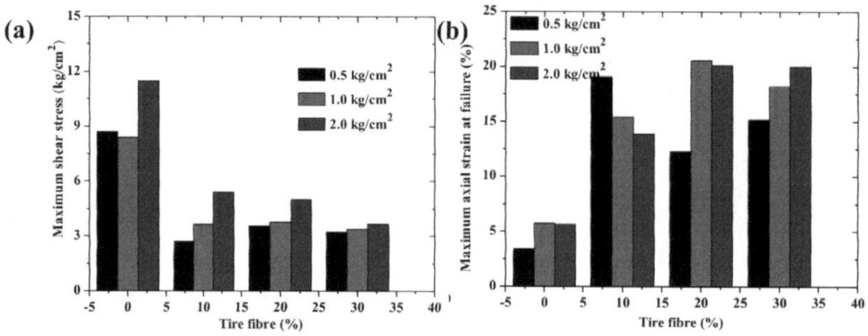

Figure 7. Variations of (a) maximum shear stress and (b) maximum axial strain with tire fibre at different confining pressures

From the Fig. 7a, it can be observed that peak shear stress decreased drastically as the tire content is increased from 0% to 10%; on further increase in TF up to 20%, the peak shear stress values are marginally increased. On further addition of TF from 20% to 30% resulted in a slight decrease in stress again. Figure 7b present the variations of maximum shear strain with the addition of tire fibre at different confining pressures. It can be observed that for higher confining pressure, the maximum axial strain achieved by the soil was found to be at 20%. Figure 8 (a-d) represents results obtained from CCT in terms of Mohr circle and failure envelope. This is very helpful to find out the shear strength parameters namely cohesion '*C*' and angle of internal friction (ϕ), which can be further utilized to estimate the shear strength of soil and soil-tire mixes. From Fig. 8 (a-d), it can be observed that ϕ decreases with increasing TF whereas, *C* decreased initially up to 10%TF, further shows increase in *C* with the addition of TF up to 30%. Table 2 shows the value of cohesion and angle of internal friction obtained at all percentage of TF. The decrease in angle of internal friction angle is estimated as 12.5%, 46%, 78%, as TF is increased from 0% to 30% respectively. While the cohesion value first decreases by 75% up to 10% TF mix and then there is a uniform increase in cohesion with increase in tire content.

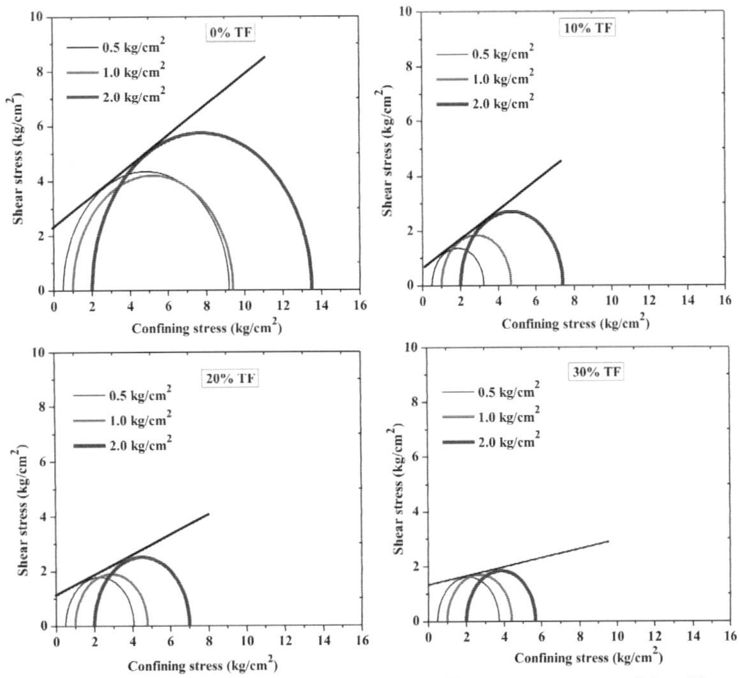

Figure 8. (a-d) Mohr circle at different percentage of tire fibre

Table 2 Value of cohesion and angle of internal friction

Tire fibre (%)	Cohesion, C (kg/cm^2)	Angle of internal friction (ϕ)
0	2.2	32
10	0.60	27
20	1.0	18
30	1.4	7

4 Energy Absorption Capacity (EAC) of Soil-tire Mixture

The Energy Absorption Capacity (EAC) of soil and soil-tire mixes was evaluated at different strain rate. EAC value was estimated calculated considering the area under the stress–strain curve of soil and soil-tire mixes (Babu and Vasudevan, 2008), which indicates the toughness of a soil sample (ASTM C1609/C1609M 2012). Figure 9 shows the typical representative diagram of stress-strain curve for obtained from UCS at different strain rate which was used to calculate EAC. Figure 10 (a, b) presents the variations in EAC with tire fibre at different shearing rate based on the results obtained from UCS and CCT tests. Based on the results obtained UCS test, Fig. 10a, it can be observed that the EAC increases with increasing percentage of tire fibre up to 10%, thereafter it decreases. It can also be noticed that EAC value has been increased by 55%, for strain rate of 1.25mm/min, when TF increased from 0%-10% whereas; the same has been increased by 67%, at strain rate of 2.5mm/min. On further increase in TF from 10% to 30%, EAC decreased by 44%, 32%, 52% at the shearing rate of 0.62mm/min, 1.25mm/min and 2.5mm/min, respectively. However, beyond 30% TF, EAC value increased by 22%, 24% and 6% at the shearing rate of 0.62mm/min, 1.25mm/min and 2.5mm/min, respectively. Figure 10b presents EAC at different soil tire mix based on the result obtained from CCT. It can be observed that the maximum EAC is achieved at 20% tire mix among all samples. It can also be noticed that the maximum EAC value increase with increasing in the confining stress. This increase in EAC was found to be 60% and 104% as confining pressure increase from 0.5 to 1 kg/cm^2 and from 0.5 to 2 kg/cm^2, respectively.

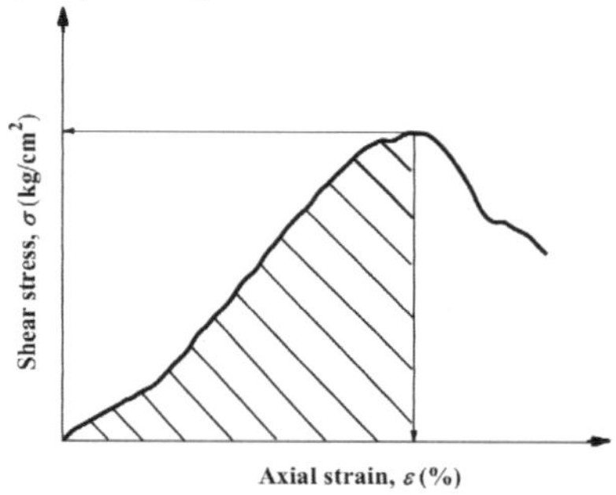

Figure 9. A typical diagram of stress-strain response to evaluate EAC using UCS and CCT test

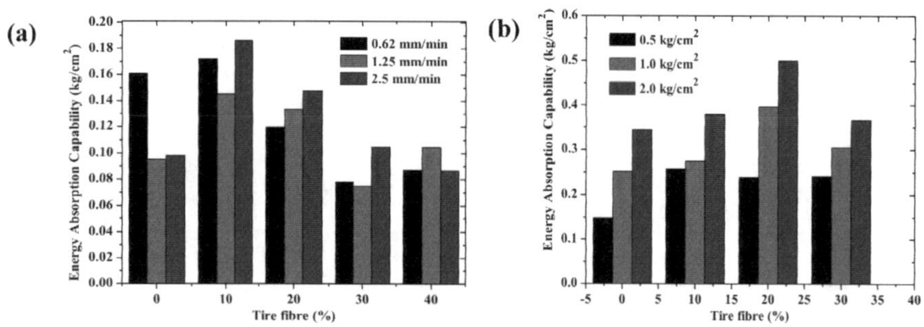

Figure 10. Variations of maximum EAC with the addition of tire fibre at different shearing rate (a) based on UCS and (b) based on CCT test

5 Conclusions

To find out the response of soil tire mixes several laboratory tests such as Compaction tests, Oedometer tests, unconfined compression tests (UCS) and confined compression tests (CCT) have been performed. Based on the results obtained from different experimental investigation the following conclusions have been drawn.

- Based on the compaction tests results, it was found that MDD values decreases with increase in TF% in mix. Further, compression index values, obtained from Oedometer tests, were also found to be decreased with increasing TF% in soil.
- Based on UCS tests it was found that, though the maximum peak stress has been reduced with the addition of tire fibre, 10 % tire fibre can be used to mix with soil since soil-tire mixes show more ductile in comparison to the other percentage of tire fibre. Based on CCT results, it can also be concluded that the mixing of 10% tire fibre with soil will provide the strength as well as more ductility in comparison to the other soil-tire fibre mixes.
- EAC was found to be higher at 10% tire fibre as well as 20% tire fibre corresponding to the maximum peak shear stress obtained from UCS and CCT, respectively. Though, the strength is a major concern for any soil engineering project, based on the results 10% of tire fibre can be recommended for the improvement ductility of silty-soil.

References

[1] Al-Tabbaa, A., Blackwell, O. and Porter, S.A., (1997). An investigation into the geotechnical propertiesof soil-tyre mixtures. *Environmental technology*, 18(8), pp.855-860

[2] Al-Tabbaa, A. and Aravinthan, T., (1998). Natural clay-shredded tire mixtures as landfill barrier materials. *Waste Management*, 18(1), pp.9-16

[3] Cokca, E. and Yilmaz, Z., (2004). Use of rubber and bentonite added fly ash as a liner material. *Wastemanagement*, 24(2), pp.153-164

[4] Özkul, Z.H. and Baykal, G., (2006). Shear strength of clay with rubber fiber inclusions. GeosyntheticsInternational, 13(5), pp.173-180.

[5] Mokhtar, M. and Chan, C.M., (2007). Effect of using cement admixed with rubber chips on the undrainedshear strength of soft soil. In National seminar on civil engineering research. Skudai: Universiti TeknologiMalaysia. http://eprints. uthm. edu. my/4150

[6] Sarvade, P.G. and Shet, P.R., (2012). Geotechnical properties of problem clay stabilized with crumb rubber powder. Bonfring International Journal of Industrial Engineering and Management Science, 2(4),pp.27-32

[7] Asadzadeh, M. and Ersizad, A., (2013). Effect of tire-chips on geotechnical properties of clayey soil. InInternational Symposium on Advances in Science and Technology, 3(2), pp. 117-120

[8] Prasad, A.S., Ravichandran, P.T., Annadurai, R. and Rajkumar, P.R.K., (2014). Study on effect of crumbrubber on behavior of soil. *International Journal of Geomatics and Geosciences*, 4(3), pp.579-584

[9] Daud, N., Yusoff, Z. and Muhammed, A., (2015). Ground improvement of problematic soft soils usingshredded waste tyre. In The Sixth Jordanian International Civil Engineering Conference (JICEC06).

[10] Signes, C.H., Garzón-Roca, J., Fernández, P.M., de la Torre, M.E.G. and Franco, R.I., (2016). Swelling potential reduction of Spanish argillaceous marlstone Facies Tap soil through the addition of crumb rubberparticles from scrap tyres. *Applied Clay Science*, 132, pp.768-773

[11] Ravichandran, P.T., Prasad, A.S., Krishnan, K.D. and Rajkumar, P.K., (2016). Effect of addition of wastetyre crumb rubber on weak soil stabilisation. Indian Journal of Science and Technology, 9(5), pp.1-5

[12] Yadav, J.S. and Tiwari, S.K., (2017). Effect of waste rubber fibres on the geotechnical properties of claystabilized with cement. *Applied Clay Science*, 149, pp.97-110.

[13] Yadav, J.S. and Tiwari, S.K., (2018). Evaluation of the strength characteristics of cement-stabilized clay–crumb rubber mixtures for its sustainable use in geotechnical applications. *Environment, development andsustainability*, 20(5), pp.1961-1985.

[14] Saberian, M. and Rahgozar, M.A., (2016). Geotechnical properties of peat soil stabilised with shreddedwaste tyre chips in combination with gypsum, lime or cement. Mires and Peat, 18(16), pp.1-16.

[15] Mukherjee, K., Mishra, A.K., (2020). Experimental and Numerical Study on Compacted Sand Bentonite-Tire Fiber Composite for Landfill Application, Lecture Notes in Civil Engineering, Advances in Computer Methods and Geomechanics, IACMAG Symposium 2019, https://doi.org/10.1007/978-981-15-0886-8.

[16] Jastrzebska and Tokarz (2021). Strength Characteristics of Clay–Rubber Waste Mixtures in Low-Frequency Cyclic Triaxial Tests, Minerals, 11, 315. https:// doi.org/10.3390/min11030315.

[17] Bowles, J.E. (1996) Foundation Analysis and Design. 5th Edition, The McGraw-Hill Companies, Inc., New York.

[18] BIS: 2720-Part VI, 1972. Determination of shrinkage factors (reaffirmed in 1995). New Delhi, India: BIS.

[19] BIS: 2720-Part VII, 1980. Determination of water content – dry density relation using light compaction (re-affirmed in 1987). New Delhi, India: BIS.

[20] BIS: 2720-Part X, 1991. Determination of unconfined compressive strength (re-affirmed in 1995). New Delhi, India: BIS.

[21] Foose, G., Benson, C., Bosscher, P., 1996. Sand reinforced with shredded waste tires. Journal of Geotechnical and Environmental Engineering 122, 760–767.

[22] Reddy, S.B., Kumar, D.P., and Krishna, A.M. (2017). Evaluation of optimum mixing ratio of sand–tyre chips mixture for geo-engineering applications. Journal of Materials in Civil Engineering, ASCE 28(2), paper no. 1335, http://dx.doi.org/10.1061/(ASCE)MT.1943-5533. 0001335.

[23] Kim, Y. T. and Kang, H. S. (2013). Effects of rubber and bottom ash inclusion on geotechnical characteristics of composite geomaterial. Marine Georesources and Geotechnology, 31(1), 71–85. doi:10.1080/ 1064119X.2012.667867

[24] Rao, G.V. and Dutta, R.K. (2006). Compressibility and strength behaviour of sand–tyre chip mixtures. Geotechnical and Geological Engineering, 24 (3), 711–724. doi:10.1007/s10706-004-4006-x.

[25] Oikonomou, N. and Mavridou, S. (2009). The use of waste tyre tyre in civil engineering work. In: J.M. Khatib, ed. Sustainability of construction materials. Cambridge, UK: Cambridge University Press

[26] Cetin, H., Fener, M. and Gunaydin, O. (2006). Geotechnical properties of tyre-cohesive clayey soil mixtures as a fill material. Engineering Geology, 88, 110–120.

[27] Srivastava, A., Pandey, S. and Rana, J. (2014). Use of shredded tyre waste in improving the geotechnical properties of expansive black cotton soil. Geomechanics and Geoengineering, 9(4), 303–311. doi:10. 1080/17486025.2014.902121.

[28] Marefat, V. and Soltani-Jigheh, H. (2011). Laboratory behaviour of clay-tire mixtures, World applied clay sciences, 13(5), 1035-1041.

[29] Soltani-Jigheh, H., Marefat, V. and Ersizad, A. (2013). Behavior of Clay-Tire Mixtures Subjected to Undrained Monotonic Loading, Soils and Rocks, São Paulo, 36(3): 283-292.

[30] Trouzine, H., Bekhiti, M. and Asroun, A., (2012). Effects of scrap tyre rubber fibre on swelling behaviourof two clayey soils in Algeria. Geosynthetics International, 19(2), pp.124-132.

IOP Conf. Series: Materials Science and Engineering 1273 (2023) 012013 doi:10.1088/1757-899X/1273/1/012013

Effect of NaOH Treated Polyethylene Terephthalate (PET) Plastic Bottle Strips on Stabilization of Soil

Sumit Kumar [1](ORCID - 0000-0001-9002-7060) and M.K.Mishra [2]

[1]Research Scholar, Dept. Of Civil Engineering, BIT Sindri, Dhanbad, Jharkhand, India.
Sumit.kumar.9@outlook.com

[1]Assistant Professor, Dept. Of Chemistry, BIT Sindri, Dhanbad, Jharkhand, India.
mkmishrabit@gmail.com

Abstract. Plastics such as polyethene bags, bottles, and so on are growing increasingly common as science progresses. The disposal of discarded material is a significant issue because most plastic wastes are non-biodegradable and unsuitable for combustion due to the emission of hazardous gases. By just using controlled compaction, soil stabilization enhances the engineering capabilities of weak sub-soils or the addition of stabilizing agents such as cement, lime, fly ash, bitumen, tar etc. But these additions have also been quite costly in recent years. This research report gives a thorough examination of the performance and effectiveness of waste plastic bottle strips in soil improvement. This waste plastic bottle is made of Polyethylene Terephthalate (PET). The size of PET plastic strips used in this study is not more than 5 mm in length and 2 mm in width. These strips were treated with Sodium Hydroxide (NaOH) to perform the proper dispersion of PET bottle fibres in soil. Although sodium hydroxide is a noncombustible solution, it is a highly reactive substance. It interacts with PET strips to exhibit changes in its properties. This work comprises replacement of soil by NaOH treated PET bottle strips in the varying percentage of 0.75%, 1%, 1.5%, 2%, 3% and 4% by dry weight of soil sample and to find out the optimum percentage mix of PET strips. The strength and geotechnical property of soil mixed with NaOH treated PET strips was evaluated and compared with the strength of soil and non-treated PET strips mix. The result shows the benefit of adding NaOH treated PET strips over non-treated strips by improving the maximum dry density (MDD), unconfined compressive strength (UCS) and California bearing ratio (CBR) value appreciably at an optimum mix of 1.5%. This soil stabilizing approach may be utilized efficiently to fulfil social and environmental problems. This exploratory inquiry would lead to the ecologically friendly management of this waste material.

[1] Research Scholar, Dept. Of Civil Engineering, BIT Sindri, Dhanbad, Jharkhand, India.

Keywords: Soil Stabilization, Polyethylene terephthalate (PET), Sodium Hydroxide (NaOH), Optimum Moisture Content (OMC), Maximum Dry Density (MDD), Unconfined Compressive Strength (UCS), California Bearing Ratio (CBR) value.

1 Introduction

Soil Stabilization is an idea to improve the physical and its onsite engineering properties employing controlled compaction with mechanical equipment's or using suitable admixture like lime, fly ash and cement or reinforcing the soil with waste materials like plastics. In past years, there has been a tremendous increase in the quantity of waste plastic; this waste plastic is most often dumped or thrown, which may disrupt the ecology and surroundings. The utilization of this substance decreases the soil's substandard qualities such as permeable and consolidating potential despite enhancing shear capacity [1]. Soil stabilization extends well beyond facilitating the usage of poor soil and minimizing the laborious operation of deficient soil regeneration [2]. There are numerous approaches of soil stabilization available these days, such as mechanical, chemical, polymers, and plastic stabilization [3]. Furthermore, it is universally acknowledged that there exist environmental issues concerning home Planet, specifically where waste disposal is a worry. These issues are caused by garbage, and among the most regularly created wastes is plastic. Plastics contribute for 12.5% of total waste produced, well with majority coming through abandoned PET bottles of water [4]. Polyethylene Terephthalate (PET) bottle are common water bottles which has rapidly gained popularity. These are often used for packaging water, carbonated beverages, liquid foodstuffs, and a range of other drinks. With their growing market, dumping had also been problematic, and breakdown of all these waste PET bottles in habitat requires a quite lengthy period. [5]. Reprocessing and salvaging plastic bottles to stabilize clay soil are advancements in the true path, and indeed the structure sector would be an ideal choice due because of its load demand. This methodology might be a realistic choice for eradicating unwanted plastic water bottles out from ecosystem [6].

Previous studies regarding soil stabilization with plastic waste as their primary additive have shown promising results. PET plastic bottle strips are used as stabilizers for two soil samples, black cotton soil and red soil. It was observed that the soil's strength had increased employing the value of CBR and MDD. It was observed that an increase in the strength was noticed by mixing 0.5 % for the black cotton soil and 0.7 % of unwanted plastic strips to red soil [7].

In previous works, it was also observed that an appreciable amount of increase in strength of soil mass when PET plastic waste was used as an additive. However, there was a dispersion problem of PET plastic waste mixed with soil, and this poor dispersion would be due to the hydrophobic nature of plastic and poor wet-ability of this plastic waste. An approach that counteract this poor dispersibility of plastic in soil, is undertaken by chemically treating these PET plastic strips that would use Sodium Hydroxide (NaOH) solution. The sodium hydroxide solution is a bleached, fragrance-free, non-volatile solution. It does not burn, but it is remarkably responsive. It interacts

pugnaciously with water, producing enough heat to ignite flammable stuff nearby. Its main benefit is that it interacts fast with water, resulting in significant compaction assistance and a greater density for the same compacting effort. [8], [9], [10].
This paper refer to an experimental examination into the effect of accumulation of NaOH treated PET strips of size (5 mm * 2 mm) on the engineering features of a natural soil sample and compares its strength properties when combined with non-treated PET strips.

2 Materials and Methodology

This study paper is made up of many experiments that are carried out in order to reach the goal and objective, which is to stabilize the soil. This was accomplished in three steps.

- The soil sample was collected on Baliapur Road in Dhanbad. The obtained soil sample was allowed to dry naturally before undergoing several tests to examine the physical qualities of the soil sample.
- Polyethylene Terephthalate (PET) strips were treated with Sodium hydroxide solution in the second phase.
- Finally, the Proctor Test, CBR, and UCS tests were done to examine the behavior of bare soil with varied percentages of sodium hydroxide-treated Polyethylene Terephthalate (PET) strips, and also the result was compared with the result obtained when soil was mixed with non-treated PET strips.

2.1 Natural Soil

In order to identify the particles size distributed in the collected soil sample, it is mandatory to observe different particle-size tests confirming IS 2720-4. By washing the soil samples over a 75-micron sieve, wet sieving distinguishes fine particles from coarse grains. The wet sieve analysis of soil showed that 63.7 % of soil passed through 75-micron sieve, and it is known that if more than 50 % of soil sample passes through, then the soil is classified as fine-grained. Further Hydrometer test has carried out, and a curve for particle size distribution was drawn.

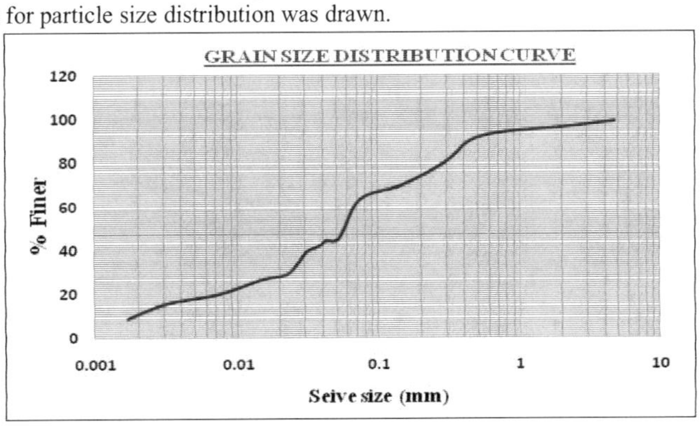

Figure 1: Particle size Distribution (PSD) curve

IOP Conf. Series: Materials Science and Engineering 1273 (2023) 012013 doi:10.1088/1757-899X/1273/1/012013

As soil has been classified as Fine grained, a series of the laboratory test was done on the natural soil sample to classify the soil type as well as natural properties,
The Plasticity Index (I_p) *of natural soil* = $W_L - W_P = 39.24 - 21.96 = 17.28$ %
Incorporating all experimental data classification of Natural soil was to be done
A-line plasticity index (I_p) = $0.73 (W_L - 20) = 14.045$ %
The Soil sample Plasticity Index (I_p) is above the A-line, indicating that it is clay, and the *liquid limit is 39.24%*, showing that it is *between 35 and 50*, indicating that it is of *intermediate compressibility*.
In conclusion, the soil categorization is **CI (clay with intermediate compressibility)** and now further stabilization methods can be carried out.
The details of all laboratory data for the natural soil sample is summarized in Table 1.

Table 1: Natural Properties of Soil

Sl. No.	Indices	Values
i.	Specific gravity (G)	2.47
ii.	Optimum Moisture Content (**OMC**)	15.39 %
iii.	Maximum Dry Density	1.71 g/cc or 16.77 KN/m^3
iv.	$\mathbf{W_L}$ (Liquid Limit)	39.24 %
v.	$\mathbf{W_P}$ (Plastic limit)	21.96 %
vi.	$\mathbf{I_P}$ (Plasticity index)	17.28 %
vii.	Unconfined Compressive Strength	3.24 kg/cm^2
viii.	Un-soaked BR	4.09 %
ix.	Soaked CBR	1.93 %

2.2 Polyethylene Terephthalate (PET)

PET is synthesized from petroleum hydrocarbons with using ethylene glycol and Terephthalate acid interactions. It is recognized as an important addition for stabilizing soil to boost the engineering aspects of subsoil to its outstanding resistance to wear, reduced coefficient of friction, and great flexural toughness. PET has the chemical formula $(C_{10}H_8O_4)_n$.

Table 2: Properties of PET plastic [11].

Behaviour Parameters	**Values**
Fiber Type	PET
Tensile strength	350 MPa
Modulus of elasticity	1800 N/mm^2
Density	1.0 g/cc
Melting point	590° c
Resistance to acidic and alkali actions	Very good

Figure 2: PET strips of size (5mm * 2mm)

2.3 Polyethylene terephthalate (PET) strip treatment, dosage, and kneading.

Seeing as PET strips are repellent, this don't really retain much amount of moisture, therefore the strips were not adequately dispersed in Soil, which is why we have been using NaOH for thorough propagation.
The following instructions were followed to prepare PET strips.

- Prepare a 5% v/v Sodium Hydroxide solution.
- Weigh some PET strips and maintain them evenly distributed for 72 hours.
- Withdraw the PET strips out from solution and cleanse with distilled water.
- Dehydrate it in an oven for 10 to 20 minutes at a temperature of 75°C to 85°C.
- The fibers is then weighed again to ensure that there has been no weight increment in the strips, and the strips are dispersed and prepared to be used as reinforcements in the native soil specimen.

At this point, treatment of PET strips have been completed and now dosing of PET strips in raw is done.

- To begin, various proportions of PET strips are introduced, namely 0.75%, 1%, 1.5%, 2%, 3%, and 4% by weight of raw soil.

Following that, treated PET strips are mixed with raw soil properly to get a uniform mixture, and water was added as per the consistency.
Different samples have been prepared for the experimental study with varying proportions of NaOH treated PET strips, given in the table below.

3 Results and Discussion

In order to explore the influence of NaOH treated PET strips on the behavior of raw soil samples, readings and results from the standard proctor test, unconfined compression test (UCS), and CBR test (wet and un-soaked conditions) were analyzed. This measured value was then compared to the strength of a soil mix made up of non-treated PET strips.

3.1 Results of Standard Proctor Tests

A conventional proctor test was carried out to replace raw soil with PET strips at varying percentages. Furthermore, the dispersion impact was investigated by comparing the replacement of raw soil with NaOH-treated PET strips and non-treated strips.
Maximum dry density increased as the percentage mix of NaOH treated PET strips in soil increased, up to an ideal mix of 1.5%. The strength of the soil in terms of maximum dry density decreased as the percentage mix of NaOH treatment increased. The increase in MDD value was 1.732 g/cc to 2.050 g/cc, a total of 18.36%.
When non-treated PET strips were substituted with raw soil, the maximum dry density increased up to a mix of 1% of the dry weight of the soil sample; nevertheless, increasing the number of strips resulted in a constant drop in value.
When NaOH-treated PET strips were ingested, the maximum dry density value was higher than when non-treated PET strips were consumed.

The highest dry density value for treated PET strips mix was 2.050 g/cc, while it was 2.024 g/cc for non-treated PET strips mix.

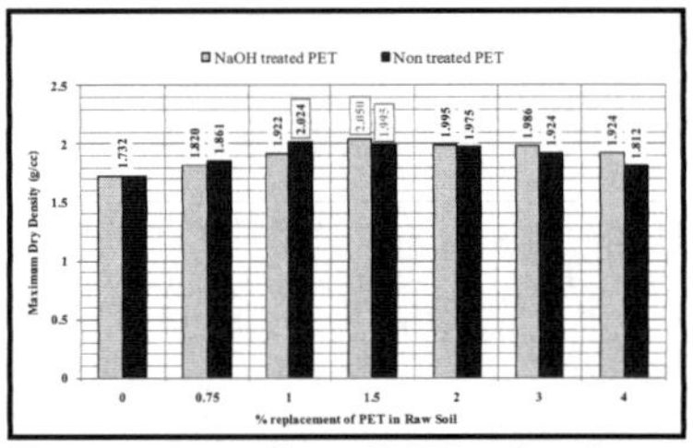

Figure 3: Comparison of MDD values for NaOH and non-treated PET strips.

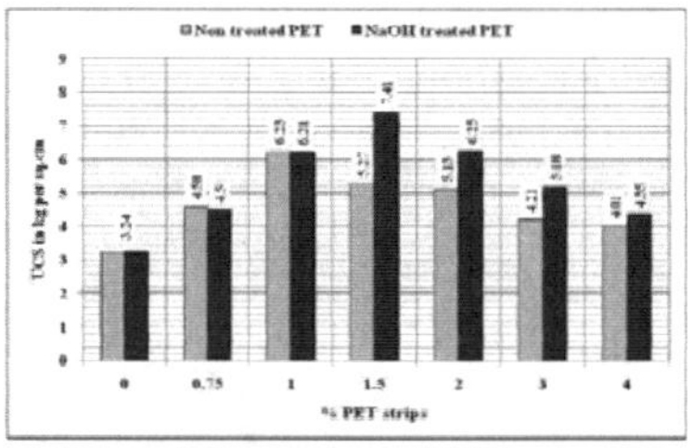

Figure 4: Comparison of UCS values for both NaOH and non-treated PET strips

When the MDD values for NaOH-treated and non-treated cases were compared, it was discovered that PET strips substituted with soil attained higher maximum dry densities at 1.5% and 1% of the dry weight of the soil sample, respectively, and then decreased.

3.2 Results of Unconfined Compressive Strength Tests

The unconfined compression test shows that natural soil has an unconfined compressive strength (UCS) of 3.24 kg/cm^2.

As the dose of treated PET strips in raw soil sample is enhanced, the soil's unconfined compressive strength (UCS) sample increases from 3.24 kg/cm^2 to 7.41 kg/cm^2 up to a replacement of 1.5% in raw soil sample. The increase was 128.70% in total.

Non-treated PET strips, on the other hand, demonstrated an increased behavior of unconfined compressive strength up to a mix of 1% of dry weight of soil sample when replaced with raw soil. However, increasing the number of strips resulted in a constant decline in value.

Table 3: UCS Test Results.

Soil (%)	PET (%)	UCS (kg/cm^2) For non-treated PET strips.	UCS (kg/cm^2) For NaOH treated PET strips.
100	0	**3.24**	3.24
99.25	0.75	4.58	4.50
99	1	**6.23**	6.21
98.5	1.5	5.27	**7.41**
98	2	5.13	6.25
97	3	4.21	5.18
96	4	4.01	4.35

The UCS strength observed in Table 3 and variation of unconfined compressive strength concerning PET strips is shown in Figure. 3.

3.3 Results of California Bearing Ratio Tests

It's an empirical test cast-off to assess the shear resistance of soil layers. This evaluation has a significant importance due to its being equally mandatory and simple to execute. Additionally, since of its extensive use throughout the world, it generates an enormous amounts of data to interpret the results. California bearing ratio is mainly a laboratory test, however the assessment is also carried out within the soil on site.
For this research analysis, a CBR test for both un-soaked and soaked conditions is carried out.

Table 4: CBR Test results for un-soaked condition

Soil (%)	PET (%)	Non-Treated PET Strips		NaOH PET Strips	
		Un-soaked CBR (%)	**Soaked CBR (%)**	**Un-soaked CBR (%)**	**Soaked CBR (%)**
100	0	4.09%	2.18%	4.09%	2.18%
99.25	0.75	5.71%	3.71%	6.01%	3.68%
98	**1**	**6.43%**	**5.67%**	7.34%	4.88%
98.5	**1.5**	5.87%	4.80%	**8.56%**	**6.86%**
98	2	5.17%	4.13%	7.51%	5.80%
97	3	5.02%	3.85%	6.85%	5.20%
96	4	4.74%	3.50%	6.77%	4.65%

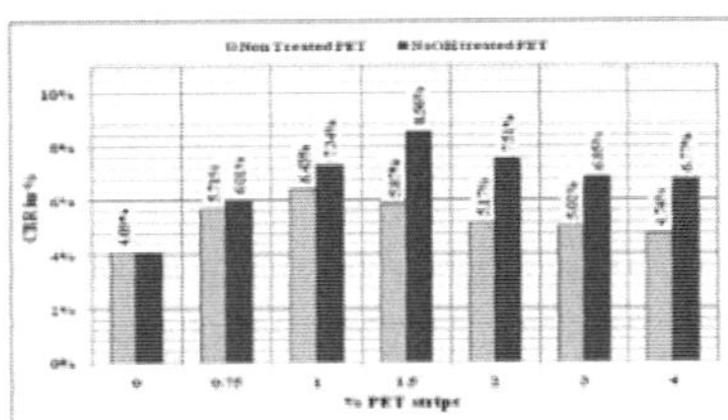

Figure 5: Comparison of Un-soaked CBR values for both NaOH and non-treated PET strips.

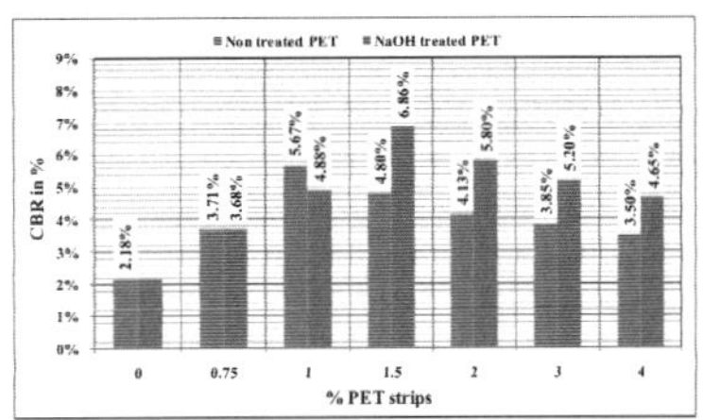

Figure 6: Comparison of soaked CBR values for both NaOH and non-treated PET strips.

The un-soaked CBR value for non-treated PET strips mixed with natural soil increased with increasing PET dose; the maximum CBR value of 6.43% was recorded at 1% mix. In contrast, the growing value of CBR for NaOH treated PET strips was observed up to a percentage mix of 1.5% with a CBR value of 8.56%; however, increasing the dose of treated PET strips in raw soil sample resulted in decreasing value..
As a result, the consumption of PET is higher for NaOH-treated PET strips, and the CBR value is higher in comparison to non-treated PET strips, as shown in Fig. 4.
The soaked CBR value for non-treated PET strips mixed with natural soil increased with increasing PET dose; the maximum CBR value of 5.67% was obtained at 1% mix. However, the increasing value of CBR for NaOH treated PET strips was seen up to a percentage mix of 1.5% with a CBR value of 6.86%, and increasing the dose of treated PET strips in raw soil sample produced a reduction in value.

IOP Conf. Series: Materials Science and Engineering 1273 (2023) 012013 doi:10.1088/1757-899X/1273/1/012013

Hence, the consumption of PET is more for NaOH treated and it also provided improved CBR value in comparison to non-treated PET strips, as shown in Fig. 5.

4 Conclusion

By observing different test results, it can be conclude that the raw soil with 1.5% of NaOH treated PET strips by weight gives optimum results:
The maximum dry density (MDD) value goes on increasing from 1.732g/cc to 2.05 g/cc up to 1.5% of PET strips replacement in the raw soil sample; after that, it keeps decreasing. Therefore, it was observed that the total increment of MDD was 18.36%.
Unconfined Compressive Strength value goes on increasing with replacement of raw soil with PET strips from 3.24 kg/cm^2 to a maximum UCS of 7.24 kg/cm^2 up to 1.5% mix; beyond this, the UCS value decreases. Finally, this increment is 128.7%, clarifying that a higher load can be born after adding it.
Soaked CBR value goes on increasing from 2.18% to 6.86% as PET strips were increased in raw soil up to a mix of 1.5%; further, the soaked CBR value decreases. Therefore, it can be used for pavement in areas with a high groundwater table.
Un-Soaked CBR value goes on increasing from 4.09% to 8.56% as PET strips were increased in raw soil up to a mix of 1.5%; further, the un-soaked CBR value goes decreasing. This clarifies that the slope of pavement can be improved for the stability of the slope and reduce the thickness of pavement in highways and railways. This would reduce the cost of construction.
The strength characteristics for non-treated PET mix with natural soil were also determined. The comparison showed that NaOH treated PET strips exhibits enhanced geotechnical properties than Non-treated PET strip mixed with natural soil.
This clearly shows a more effective dispersion of PET strips on the soil when treated with NaOH.

• References

1. Seco, A., Ramírez, F., Miqueleiz, L. and García, B. (2011). "Stabilization of Expansive Soils for Use in Construction". Applied Clay Science, 51, 348-352.
2. Arora, K.R. (2005), "Soil Mechanics and Foundation Engineering". Standard Publishers Distributors, New York.
3. Habiba Afrin (2017), A Review on Different Types Soil Stabilization Techniques, International Journal of Transportation Engineering and Technology. Volume 3, Issue 2, June 2017, pp. 19-24. doi: 10.11648/j.ijtet.20170302.12.
4. Ramadevi K, Manju R. (2012)."Experimental investigation on the properties of concrete with plastic PET (bottle) fibres as fine aggregates". International journal of emerging technology and advanced engineering, 2(6) 42-46.
5. Akçaözoğlu, S., Atiş, C.D. and Akçaözoğlu, K. (2010) "An Investigation on the Use of Shredded Waste PET Bottles as Aggregate in Lightweight Concrete". Waste Management, 30, 285-290.
6. Saini, K., Chaudhary, V., Bisnohi, A., Agarwal, H., Ram, M. and Saraswat, S. (2016). "Effect on Strength Properties of Concrete by Using Waste Wood Powder as Partial Replacement of Cement". SSRG International Journal of Civil Engineering, 3, 172-176.

IOP Conf. Series: Materials Science and Engineering 1273 (2023) 012013 doi:10.1088/1757-899X/1273/1/012013

7. Harish, C. and Ashwini, H. M. (2015) "Stabilization of soil by using plastic bottle strips as a stabilizer", Int. Res. J. of Engineering and Technology. Vol.4, pp 1877.
8. Gogo (2016). "Evaluated the stabilization of soil from Ghana for construction purpose by mixing and curing with various additives; NaOH, Na_2Co_3, Ca $(OH)_2$". IJSRD - International Journal for Scientific Research & Development| Vol. 3, Issue 11, 2016 | ISSN (online): 2321-0613. (2009).
9. Olaniyan, O.S., Olaoye, R.A, Okeyinka, O.M, and Olaniyan, D.B. (2011) "Soil Stabilization Techniques Using Sodium Hydroxide Additives". International Journal of Civil & Environmental Engineering IJCEE-IJENS Vol: 11 No: 06.
10. Olaniyan.O. S, Olaoye.R. A (2011). "Soil stabilization techniques using sodium hydroxide additives". International Journal of Civil and Environmental Engineering (IJCEEIJENS) Vol.11, pp 19-22, December 2011.
11. ASTM D638. Test method for tensile properties of Plastics. (2007).

Effect of subterranean levels on the dynamic response of RC-MRF buildings

Z L Tadesse[1, 3], **H K Padavala**[1], **and V R P Koteswara**[2]

[1] Department of Civil Engineering, National Institute of Technology, Warangal, Telangana 506 004, India [2] Department of Civil Engineering, Vardhaman College of Engineering, Hyderabad, Hyderabad, Telangana 501 218, India
[3] Faculty of Civil Engineering, Arba Minch Institute of Technology, Arba Minch, SNNPR, Ethiopia

E-mail: zeleke@student.nitw.ac.in

Abstract. The dynamic response of reinforced concrete moment resisting frame (RC-MRF) building structures is mostly altered by the substructure system such as the type of foundations and features of subterranean levels, as well as subsurface soil conditions. The majority of low- to high-rise buildings, in urban areas, are designed with one or multi-level subterranean levels for parking and other services. However, many designers overlooked incorporating these subterranean levels and soils in the seismic resistance analysis assuming that the building is fixed at the ground surface. The present study assessed the effect of the subterranean levels on the dynamic response of RC-MRF buildings. Several two-dimensional (2D) nonlinear finite element seismic response analyses were carried out in ABAQUS software for medium-rise buildings without and with one to multi-level subterranean levels resting on a deep homogenous soft soil profile under the effect of strong seismic input motion. It was observed that the subterranean levels considerably alter the dynamic characteristics and demands of the building structures. Hence, explicitly incorporating the subterranean components and foundation soil during the seismic resistance design of building structures is important.

Keywords: Soil-structure interaction · Subterranean levels · Seismic response · Finite element method.

1. Introduction

All of the building structures comprise some type of substructure system such as different foundation types and subterranean levels, which are in direct contact with soil. Designing low-rise to high-rise buildings having one or multiple subterranean levels with foundations is the best solution in urban areas to provide additional spaces for parking and other services. To evaluate the accurate dynamic response of the superstructure it is needed to consider the substructure system and the soil supporting structure in the soil-structure interaction (SSI) analysis. Due to the lack of detailed guidelines in the local design codes, overlooking or incorporating the subterranean levels in the seismic resistance design of building structures is still a controversial issue [1–2]. When a building rests on a shallow foundation and is subjected to lateral seismic loads, it is reasonable to crop the superstructure and analyze it as a fixed base structure at ground level with free-field ground motion excitation since free-field motion is approximately equal to foundation input motion. However, the presence of one or multiple subterranean levels and deeply embedded foundations alters the propagating shear wave velocity and leads to refraction and reflection of the incidence seismic wave. This causes the ground motion at the foundation level to deviate from the free-field motion, often referred to as kinematic interaction. Additionally, the inertial interaction between the subsoil deposite and substructure system alters the seismic structural response. Hence, to assess the effect of the subterranean levels incorporation, the SSI effect should be taken into account [3–10]. Applying various foundation modeling approaches, the impact of foundation modeling on the response analysis of building with subterranean levels was assessed by [6]. It was proved that SSI can alter the response of buildings with subterranean levels, whereas fixing the structure at the base level provides poor results. The seismic performance of three-dimensional low to medium-rise moment-resisting frame steel buildings with multiple subterranean levels was also investigated by [7] using the substructure approach. It was found that SSI impacts are considerable for buildings without subterranean levels, and gradually decrease as the number of subterranean levels increases. More recently, the effect of dynamic SSI in building structure with subterranean levels was evaluated by [11] using physical reduced-scale modes of a building under different configuration of superstructure and subterranean levels. The results of the experimental investigations, different combinations of impacts consistently reduce the effective base motion, indicating that it would be a conservative assumption to ignore the effects of dynamic SSI.

The SSI characteristics of structures supported by shallow and deep foundations have received the majority of researcher's attention in the last few decades. However, fewer studies [2] have been presented in the literature on buildings with subterranean levels that incorporate seismic SSI utilizing an experimental and simplified three-step method (substructure approach). In the analysis of the seismic SSI response of any particular problem, numerical techniques are highly capable of incorporating different combinations of effects and nonlinearity resulting from material behaviour and changes in geometry, material nonhomogeneous properties, soil-structure interface conditions, radiation and material damping changes, as well as strong ground motion [3].

In the present study, a seismic response of a soil-structure system (SSS) was analysed using finite element method. To asssess the effect of subterranean level on the seismic response of the RC-MRF building, soft soil subsurface condition has been taken into consideration as the hypothetical site soil. The 2D finite element model of the entire SSS was simulated using ABAQUS v. 6.14 software. The simulated buildings with subterranean levels are equivalent 2D 10 and 15-story buildings comprising moment-resisting frames that representing medium-rise buildings. The earthquake acceleration time history is selected from real ground motion records that representing strong free-field seismic motions. The primary goal of this investigation is to evaluate the effect of subterranean levels and the response of the SSS in soft soil profile under strong ground motion. Moreover, this study mainly focused on the relative lateral story level displacement, lateral inter-storey drifts, roof level lateral displacement, as well as foundation level lateral motion of the building rather than determining the seismic demand forces in the structural elements, and consequently, the structural components of the RC-MRF building are assumed to undergo nonlinear inelastic behavior.

2. System under investigation

Ten and fifteen-story RC-MRF buildings regularly in plan (Figure 1(a)) and elevation (Figure 1(b)) representing the conventional medium-rise type of buildings are adopted as a reference for this study. The modeling description of the superstructures is presented in Table 1. The columns of the building are supported by an embedded mat foundation with one, three, and five basement floor systems (Figure 2), resting on a homogenous soft soil profile underline by rigid bedrock (Table 2). The building was designed according to the Indian standard IS 456 [12] in a relatively high-risk earthquake-prone zone-V response spectra.

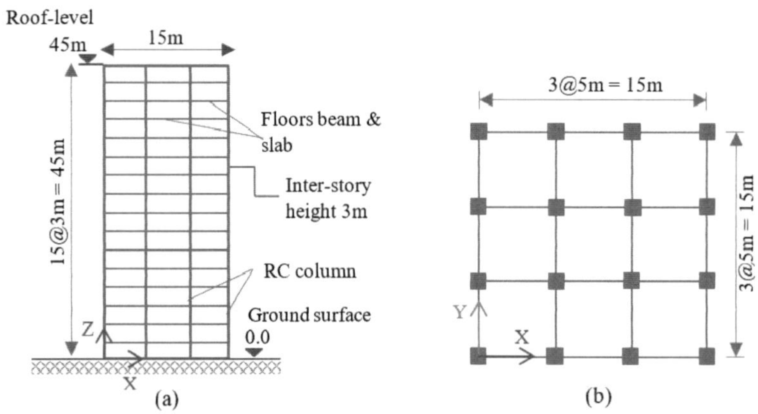

Figure 1. The layout of the fifteen-story RC-MRF building: (a) superstructure elevation view, and (b) standard floor plan of the building.

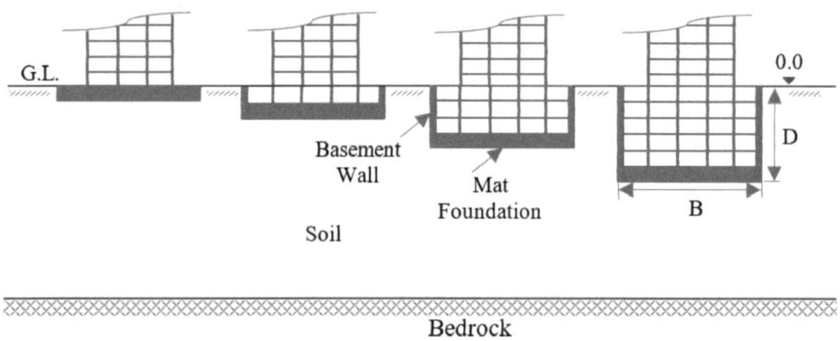

Figure 2. Schematics of the building subterranean levels with 0BS, 1BS, 3BS, and 5BS.

Table 1. Modeling description of the rigid base-superstructure.

Reference name (ID)	Number of stories	Story height (m)	Number of bays (m)	Bay width (m)	Total width (m)	Total height (m)
10S	10	3	3	5	15	30
15S	15	3	3	5	15	45

In accordance with standard engineering design procedures (IS 456 [12], Bowles [13]), the mat foundation was designed to sustain the static and dynamic loads of the structure, to meet the requirements for maximum settlement and bearing capacity, whereas the integrated basement walls were intended to support the lateral earth pressure, the shear and bending moment.

Table 2. Engineering properties of the soil considered in this study [14–15].

Soil type	USCS	VS,30 (m/s)	Gmax (MPa)	ρ (kg/m³)	E (MPa)	ν	SPT	PI (%)	C' (kN/m²)	ϕ' (°)
E[a] / III[b]	CL	150	33.1	1470	91.7	0.40	6	15	20	12

[a] ASCE 7-10 [16].
[b] IS 1893:2000 [17].
USCS – unified soil classification system.

In the present study, a seismic record of Christchurch, New Zealand (Table 3 and Figure 3) has been considered for dynamic response analyses. This earthquake ground motion was incorporated into the numerical simulation while carrying out a nonlinear time history analysis. The chosen earthquake ground motion is free-field motion.

Table 3. Details of earthquake ground motion data considered for the present study.

Earthquake Event	Year	Mw (R)	PGA (g)	Duration (sec)
Christchurch, New Zealand	2011	6.2	0.371	49.985

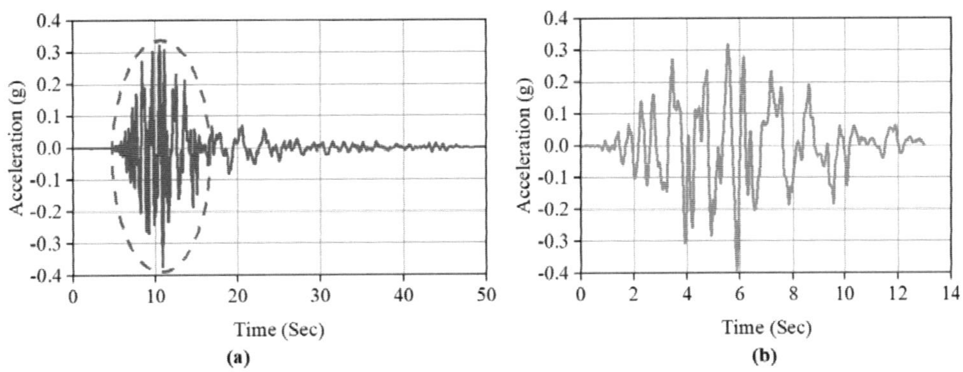

Figure 3. The 2011 Christchurch - New Zealand earthquake (a) free-field acceleration time history, and (b) part of ground motion considered for the present study.

3. Numerical model

The numerical modeling of the SSS was performed with the finite element method by the use of ABAQUS software [18]. The components of structural members and the subsoil deposite are simulated to behave nonlinearly. A Rayleigh damping of 5% is considered for the selected two modes of each soil-structure integrated system model. In order to evaluate the effect of subterranean levels on the response of RC-MRF building a direct approach was employed in ABAQUS software, in which the superstructure, mat foundation with subterranean levels, and soil are explicitly simulated using a detailed 2D finite element model (Figure 5). The structural frame elements were modelled using standard 1D beam elements, whereas the foundation element was simulated with 2D shell elements, assuming initially a linear elastic isotropic behaviour for all elements of the structural system. The foundation and subterranean levels are assumed to be bounded perfectly with the soil without interface elements.

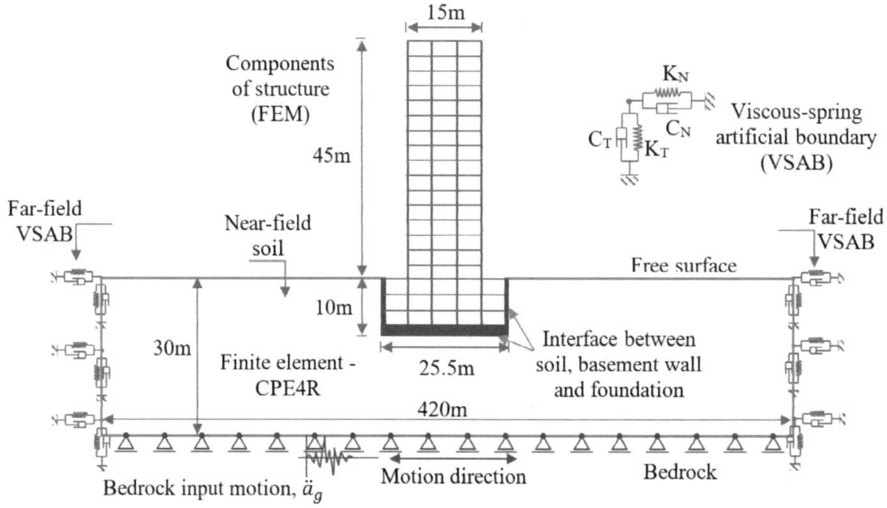

Figure 4. The direct approach configuration employed in this study for modeling the integrated SSS of a 15-stories building with subterranean levels.

The 2D numerical model of the subsoil profile was simulated as nonlinear elastic-plastic material utilizing the Mohr-Coulomb failure criterion. To avoid multiple reflections during the dynamic response analyses, to reduce a special dimension of the soil medium, as well as to optimize the computational effort, the unbounded soil domain was modeled with a viscous-spring artificial boundary consisting of elastic springs and dashpots according to [19–21](Figure 4). Kinematic constraints were employed along the lateral boundaries to simulate acceptable dynamic boundary conditions and avoid the flexure mode of vibration of soil mass [22]. These constraints prevent the soil mass from lateral spread due to gravitational loads and allow the shear behaviour between neighbouring soil layers.

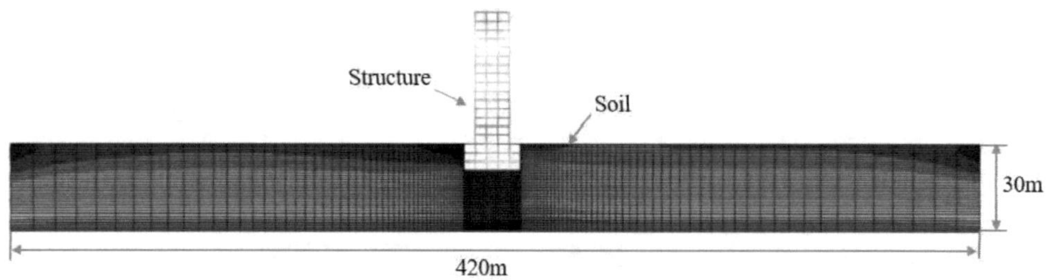

Figure 5. The integrated SSS 2D finite element model.

As suggested by [23], the bottom of the soil medium (bedrock) was simulated using a rigid boundary condition in the seismic response analysis, while the excitation motion was applied at the bedrock levelsand propagating upwards through the entire model. Through sensitivity analysis, a soil domain with anextent of 420m in the longitudinal direction was considered. As observed by some numerical and experimental tests [24], and also suggested by modern seismic resistance design codes (FEMA 450 [25],ATC-40 [26]) the most local site amplification occurs within the top 30m of the soil deposit; thus, the vertical extent (bedrock depth) of the finite element soil models were limited to 30m in this study. In most numerical dynamic SSI analyses the size of elements mainly depends on the geometry of the embedded component of the structural system and loading conditions (static or dynamic). The element sizes are often refined near the targeted area to take into account the severe stress gradients and plasticity encountered in the soil medium, whereas a gradual transition to a coarser element considers far from the targeted area in lateral directions of the soil medium. On the other hand, to permit for a uniform distribution of vertically propagating shear waves, the size of elements is kept constant in the vertical direction. Therefore, maximum size, h_{max}, of the elements was estimated using equation (1) in this study.

$$h_{max} \leq \frac{1}{a} \times \lambda \tag{1}$$

where, $\lambda = V_S / f_{max}$, in which V_S = the soil smallest shear wave velocity of interest, f_{max} = maximum frequency of interest, and a = 5, 8, and 10 as per [27], [28], and [29], respectively.

4. Results and Discussions

A 2D numerical analyses have been conducted to assess the effect of the subterranean levels on the lateral displacement behavior of RC-MRF buildings under strong ground motion. The horizontal displacements of the superstructure are evaluated at various heights in order to better comprehend the effect of the subterranean levels. The lateral inter-story displacement, the roof's level lateral displacement, inter-story drifts, and foundation level motion are also determined. The results are compared when the buildings are given 0, 1, 3, and 5 subterranean levels, which are represented by BS0, BS1, BS3, and BS5, respectively.

The dynamic responses of the 10 and 15-storey RC-MRF buildings without and with subterranean levels are presented from Figure 6 through Figure 11. To create sufficient distance between nearby structures during a seismic event, story displacement is a crucial component for preventing adjoining structures from colliding with one another. Figure 6 and Figure 7 presents the maximum lateral inter-story displacements of the superstructures. The deflection profile varies based on the subterranean levels. It has been observed that the effect of subterranean levels considerably increases for the buildings with one and three subterranean levels; on the contrary, the effect reduces for buildings with five subterranean levels.

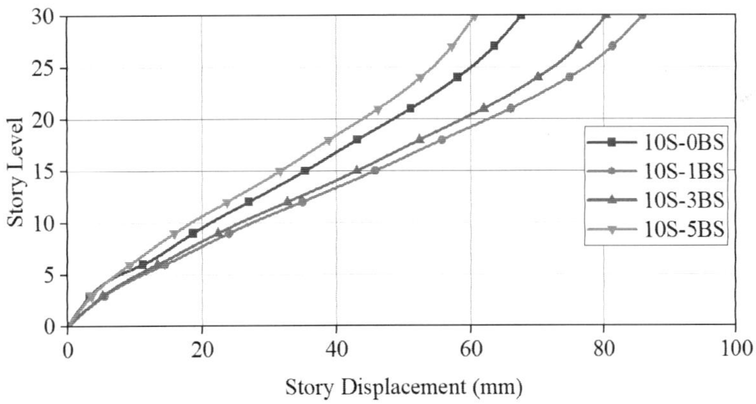

Figure 6. Maximum lateral inter-story displacement for 10-story building.

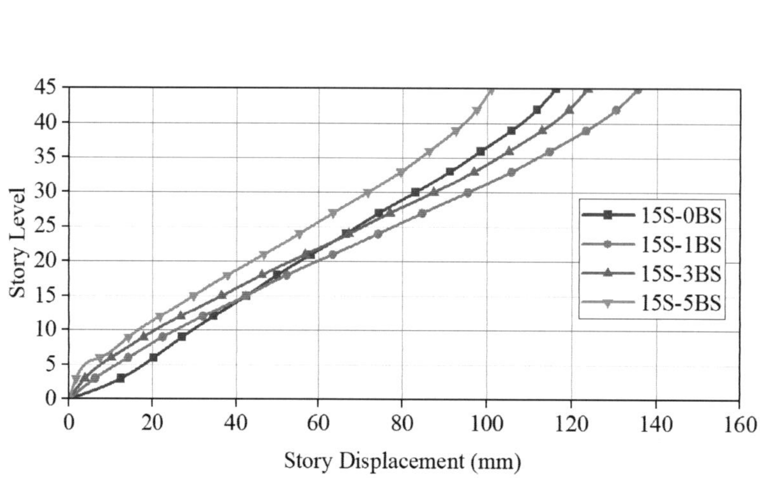

Figure 7. Maximum lateral inter-story displacement for 15-story building.

Figure 8 also shows specifically the variations of the maximum roof level lateral displacement of the two medium-rise buildings with various subterranean levels. From numerical response analysis, the roof level lateral displacement increases as the subterranean level increase from zero to three, whereas it decreases for buildings having five subterranean levels. Moreover, large roof-level lateral displacement is observed in the case of a 15-story building compared to a 10-story one.

In the assessment of the effects of lateral loads on vertical members in stability analysis, story drift is one of the crucial factors. Figure 9 and Figure 10 shows the inter-story drift in buildings with subterranean levels compared with buildings without ones. As observed in the previous results, the same variations are observed in the inter-story drift values.

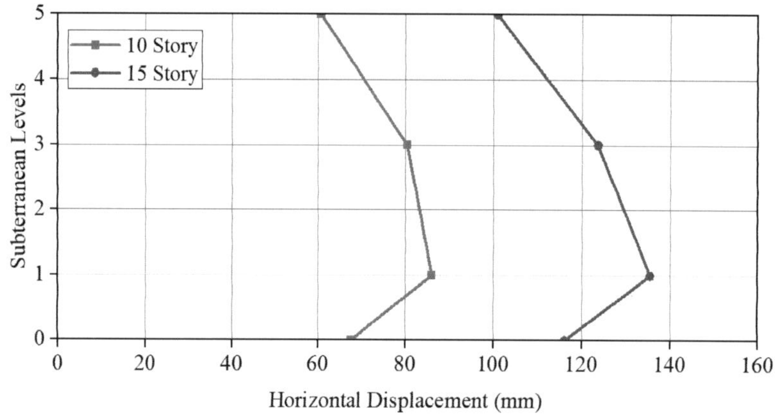

Figure 8. Maximum roof level lateral displacement for 10 and 15-story buildings with different subterranean levels.

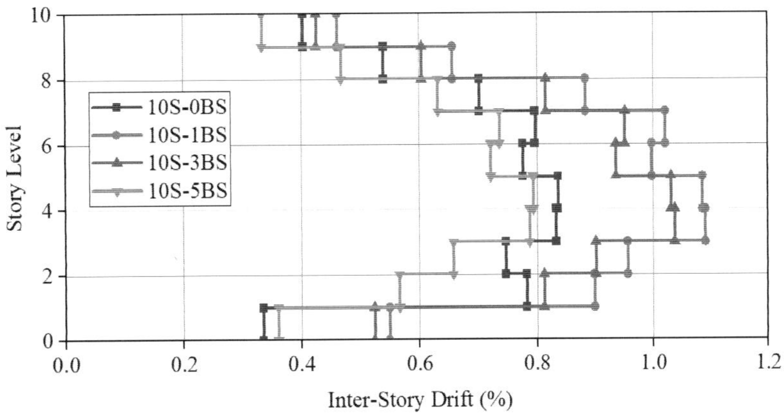

Figure 9. Inter-story drift for 10-story building.

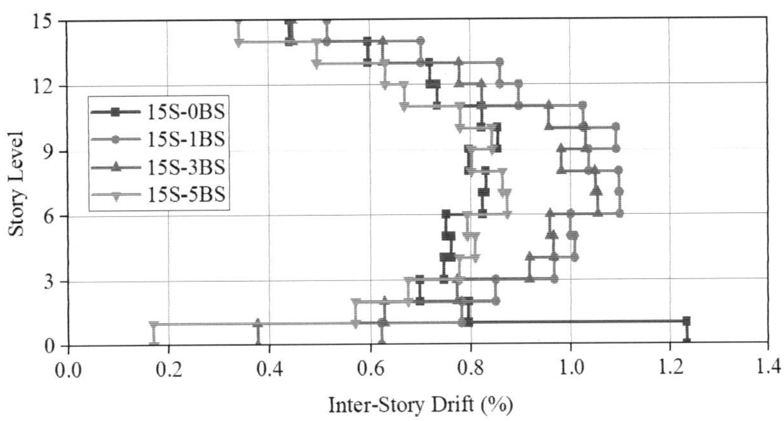

Figure 10. Inter-story drift for 15-story building.

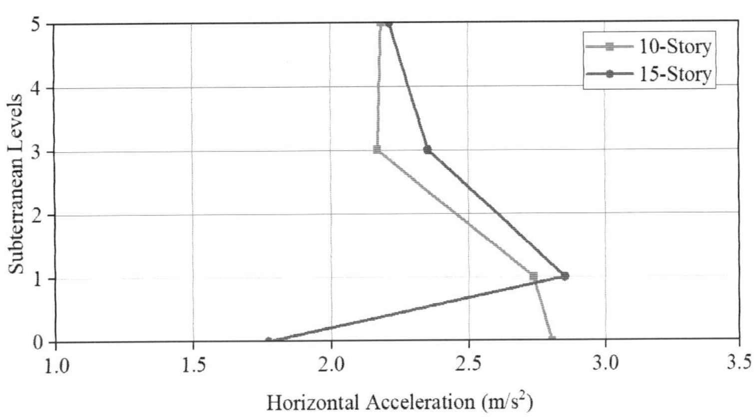

Figure 11. Maximum subterranean or foundation level horizontal acceleration.

Seismic response analysis of building structures supported on compliant foundations often uses a substructure method, in which the underlying and surrounding soil is not explicitly incorporated, but rather are replaced by foundation impedance (springs and dashpots), and the analyses are carried out under the excitation of foundation level input ground motion. In the case of building structures with one or more levels of embedded subterranean, the characteristics and magnitude of the foundation level motions may be affected by the presence of these modules. Figure 11 presents the variation of the maximum horizontal acceleration at the level of the foundation with different subterranean levels. According to the findings, the foundation level motion is significantly altered by the subterranean levels with various embedment depth.

5. Conclusions

This paper evaluated the effect of incorporating one or several subterranean levels on the dynamic response of RC-MRF buildings using the 2D numerical method. This was carried out utilizing comparisons of the nonlinear dynamic response analysis results of the two medium-rise buildings with and without subterranean levels resting in soft soil under strong earthquake ground motion. The results appeared to reveal that as the number of subterranean levels increased from zero (i.e., without subterranean level) to one or three, the lateral inter-story displacement, roof level displacement, and lateral inter-story drifts increased. On the contrary, the dynamic response characteristics of the building structure considerably reduced as the number of subterranean levels increased to five. Furthermore, when the height of the superstructure increased, the lateral inter-story displacement and roof-level displacement value significantly increased simultaneously. Moreover, the foundation level motions are also altered due to the presence subterranean levels. This assessment revealed that the structural system dynamic response is the result of an interaction between the superstructures, substructure systems, and lateral as well as foundation soil. Hence, disregarding the subterranean levels during the designing of building structures might lead to underestimating or overestimating the lateral stiffness of the building's lateral load-resisting systems.

References

[1] G. Saad, F. Saddik, and S. Najjar, "Impact of Soil Structure Interaction on the Seismic Design of Reinforced Concrete Buildings with Underground Stories," *Proc. Fifthteenth World Conf. Earthq. Eng.*, p. 4935, 2012.

[2] Z. L. Tadesse, H. K. Padavala, and V. R. P. Koteswara, "Seismic response assessment of building structures with underground stories: a state-of-the-art review," *Innov. Infrastruct. Solut.*, vol. 7, no. 6, pp. 1–18, 2022, doi: 10.1007/s41062-022-00942-5.

[3] S. C. Dutta and R. Roy, "A critical review on idealization and modeling for interaction among soil-foundation-structure system," *Comput. Struct.*, vol. 80, no. 20–21, pp. 1579–1594, 2002, doi: 10.1016/S0045-7949(02)00115-3.

[4] S. C. Dutta, K. Bhattacharya, and R. Roy, "Response of low-rise buildings under seismic ground excitation incorporating soil-structure interaction," *Soil Dyn. Earthq. Eng.*, vol. 24, no. 12, pp. 893–914, 2004, doi: 10.1016/j.soildyn.2004.07.001.

[5] H. Shakib, "Evaluation of dynamic eccentricity by considering soil-structure interaction: A proposal for seismic design codes," *Soil Dyn. Earthq. Eng.*, vol. 24, no. 5, pp. 369–378, 2004, doi: 10.1016/j.soildyn.2004.01.003.

[6] F. Naeim, S. Tileylioglu, A. Alimoradi, and J. P. Stewart, "Impact of foundation modeling on the accuracy of response analysis for a tall building," *Semin. Util. Strong Motion Data*, pp. 19–55, 2008.

[7] H. El Ganainy and M. H. El Naggar, "Seismic performance of three-dimensional frame structures with underground stories," *Soil Dyn. Earthq. Eng.*, vol. 29, no. 9, pp. 1249–1261, 2009, doi:

10.1016/j.soildyn.2009.02.003.
[8] P. Raychowdhury, "Seismic response of low-rise steel moment-resisting frame (SMRF) buildings incorporating nonlinear soil-structure interaction (SSI)," *Eng. Struct.*, vol. 33, no. 3, pp. 958–967, 2011, doi: 10.1016/j.engstruct.2010.12.017.
[9] H. R. Tabatabaiefar and A. Massumi, "A simplified method to determine seismic responses of reinforced concrete moment resisting building frames under influence of soil-structure interaction," *Soil Dyn. Earthq. Eng.*, vol. 30, no. 11, pp. 1259–1267, 2010, doi: 10.1016/j.soildyn.2010.05.008.
[10] N. Çağlar, S. Sert, and A. H. Serdar, "Basement-Storey Effect on the Seismic Response of RC Buildings on Soft Surface Soil," *Arab. J. Sci. Eng.*, vol. 46, no. 11, pp. 11291–11302, 2021, doi: 10.1007/s13369-021-05714-4.
[11] H. Segaline, E. Sáez, and J. Ubilla, "Evaluation of dynamic soil-structure interaction effects in buildings with underground stories using 1 g physical experimentation in a transparent shear laminar box," *Eng. Struct.*, vol. 266, no. 114645, 2022, doi: 10.1016/j.engstruct.2022.114645.
[12] IS 456, "Code of practice for plain and reinforced concrete (third revision)," *Bur. Indian Stand. New Delhi*, pp. 1–114, 2000.
[13] J. E. Bowles, *Foundation Analysis and Design*. 1997.
[14] A. Massumi and H. R. Tabatabaiefar, "Effects of Soil-Structure Interaction on Seismic Behaviour of Ductile Reinforced Concrete Moment Resisting Frames," *Proc. World Hous. Congr. Afford. Qual. Hous.*, 2007.
[15] S. H. Reza Tabatabaiefar, B. Fatahi, and B. Samali, "Seismic Behavior of Building Frames Considering Dynamic Soil-Structure Interaction," *Int. J. Geomech.*, vol. 13, no. 4, pp. 409–420, 2013, doi: 10.1061/(asce)gm.1943-5622.0000231.
[16] ASCE/SEI 7-10, *Minimum Design Loads for Buildings and Other Structures*. 2010.
[17] IS 1893 (Part 1), "Criteria for Earthquake resistant design of structures,Part 1:General Provisions and buildings," *Bur. Indian Stand. New Delhi*, vol. 1893, pp. 1–44, 2016.
[18] A. Duval *et al.*, "Abaqus/CAE 6.14 User's Manual," *Dassault Systémes Inc. Provid. RI, USA*, vol. IV, pp. 1–6, 2014.
[19] J. Lysmer and R. L. Kuhlemeyer, "Finite dynamic model for infinite media," *J. Eng. Mech. Div.*, vol. 95, no. 4, pp. 859–877, 1969, doi: 10.1061/jmcea3.0001339.
[20] C. Song and J. P. Wolf, "Dynamic Stiffness of Unbounded Medium Based on Damping-Solvent Extraction," *Earthq. Eng. Struct. Dyn.*, vol. 23, no. 2, pp. 169–181, 1994, [Online]. Available: https://doi.org/10.1002/eqe.4290230205
[21] D. H. Chen, C. Bin Du, J. W. Yuan, and Y. W. Hong, "An investigation into the influence of damping on the earthquake response analysis of a high arch dam," *J. Earthq. Eng.*, vol. 16, pp. 329–349, 2012, doi: 10.1080/13632469.2011.638697.
[22] A. G. Sextos, G. D. Manolis, A. Athanasiou, and N. Ioannidis, "Seismically induced uplift effects on nuclear power plants. Part 1: Containment building rocking spectra," *Nucl. Eng. Des.*, vol. 318, pp. 276–287, 2017, doi: 10.1016/j.nucengdes.2016.12.035.
[23] B. Fatahi and S. H. R. Tabatabaiefar, "Fully Nonlinear versus Equivalent Linear Computation Method for Seismic Analysis of Midrise Buildings on Soft Soils," *Int. J. Geomech.*, vol. 14, no. 4, p. 04014016, 2014, doi: 10.1061/(asce)gm.1943-5622.0000354.
[24] M. H. Rayhani and M. H. El Naggar, "Numerical Modeling of Seismic Response of Rigid Foundation on Soft Soil," *Int. J. Geomech.*, vol. 8, no. 6, pp. 336–346, 2008, doi: 10.1061/(asce)1532-3641(2008)8:6(336).
[25] FEMA 450, "NEHRP Recommended Provisions for Seismic Regulations for New Buildings and Other Structures," *Part 1*, p. 338, 2003.
[26] ATC 40, "Seismic evaluation and retrofit of concrete buildings (Report No. ATC-40)," *Appl. Technol. Counc. Redw. City, California, USA*, 1996.
[27] ASCE/SEI 4-98, "Seismic analysis of safety-related nuclear structures," 1998.
[28] R. L. Kuhlemeyer and J. Lysmer, "Finite Element Method Accuracy for Wave Propagation

Problems," *Journal of the Soil Mechanics and Foundations Division*, vol. 99, no. 5. pp. 421–427, 1973. doi: 10.1061/jsfeaq.0001885.

[29] ASCE/SEI 4-16, *Seismic analysis of safety-related nuclear structures*, vol. 4. 2017. doi: 10.1061/9780784413937.

Assessment of Lateral Load Capacity of Single Pile at Bettiah Site: A Parametric Study

Prabhakar Kumar[1], Shiva Shankar Choudhary[2], and Shiv Shankar Kumar[3]

[1,2,3]National Institute of Technology, Patna, India
[1]prabhakark.phd20.ce@nitp.ac.in, [2]shiva@nitp.ac.in, [3]k.shiv.ce@nitp.ac.in

Abstract. Pile foundations are widely used and accepted approach to transfer superstructure load to a deeper and stronger stratum. This work has been focused on the assessment of the effect of different parameters associated with piles such as pile diameter, pile length, grade of concrete and pile head condition (i.e., free or fixed) on the lateral load capacity. To evaluate the lateral load capacity considering the aforementioned parameters 'IS Code' and 'Matlock and Reese' methodologies have been utilized. The soil exploration data has been collected from ten different boreholes near to the railway bridge at Bettiah site, Bihar. The results of lateral pile load at different boreholes reflects that the lateral load capacity of pile significantly increased with the increase of pile diameters and grades of concrete. The lateral load capacity of pile was increased approximately by 13%, from both IS Code and Matlock and Reese methods, when the grade of concrete was increased from M25 to M40. It was also found that the condition of pile head also plays a major role in lateral load capacity. The lateral load capacity of pile obtained from IS code method under fixed head condition was found to be higher as compared to free head condition. It has also been observed that the lower values of ultimate lateral load capacity for fixed head and lower values for free head condition was obtained by IS Code as compared to Matlock and Reese method.

Keywords: Pile Foundations, Lateral Load Capacity, IS Code, Matlock and Reese Method, Fixed and Free Head.

1 Introduction

Pile foundations are widely used and accepted approach to transfer superstructure load to a deeper and stronger stratum in most of the pile supported structure such as bridges, tall fireplaces, TV towers, tall structures, high holding dividers and seaward constructions. Piles are utilized if superstructure's load is significantly high or settlement due to the heavy load is considerably high. Pile embedded into the soil under such kind of constructions always subjected to the lateral load due to the wind load and traffic load, which lead to translation, rotation, or bending of the pile. Winkler [1] introduced the concept of soil-pile approximation to analyze pile under lateral load, through spring but it limits the coupling effect of spring and continuity of the soil. This limitation leads to the development of p-y curve method which utilizes finite difference method to model the soil resistance [2]. P-y curve approach gives a reasonable result and that is why the study of pile behaviour under lateral action is widely adopted in the software program like LPILE and FBPIER [3]. Present study incorporates the use of subgrade reaction approach through P-y curve method. The behaviour of a laterally loaded pile is kindred to a beam on an elastic foundation subjected to transverse loads [1]. For a vertical pile, horizontal loads and the associated moments act at or above the ground surface only. Lateral load is mobilized by lateral resistance of nearby soil; thus, lateral loading capacity depends on soil properties. One of the requirements for a satisfactory piles foundation under such structures is that the maximum horizontal movement of the pile should not exceed 1% of pile diameter [4]. The detailed investigations have been carried out by several researchers for single vertical pile subjected to horizontal load [5,6], utilizing the concept of modulus of subgrade reaction [1], which further approximates the soil medium as a series of non-linear lateral independent elastic springs that are infinitely closely spaced and represents stiffness of soil by decreased length of the spring down the pile (Fig. 1). At shallow depth, the spacing between pile and spring represents the displacement of soil under the action of repeated loading and the blocks of varying size attached with springs take care of reaction, which increases with deflection and then reaches a maximum value (Fig. 1). Further, a generalized solution to calculate the lateral load for vertical piles were given by Matlock and Reese (M&R) [7]. An extensive theoretical and experimental investigations have been carried out on single vertical piles subjected to lateral loads and the results were presented in the form of bending moment and deflection [8, 9, 10]. An attempt has been made to study the behaviour of laterally loaded piles resting at various borehole locations using the Indian Standard (IS) method [11] and p-y curve approach [7], and results are represented in the form of Ultimate Lateral Loading Capacity (ULLC).

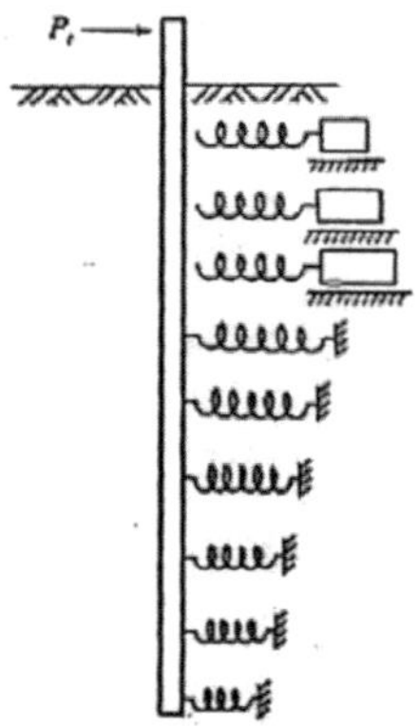

Figure 1. Approximation of laterally loaded pile-soil system [1]

2 Study Area

The site is situated 3 km North-west of Bettiah city in the West-Champaran district of Bihar state of India (latitude 26.815483° N and Longitude 84.502715° E) as presented in Fig. 2. The Piles were installed at the proposed site for the construction of Railway over-bridge Bettiah-Kumar Bagh Station or Road and the work comes under the Samastipur Railway division.

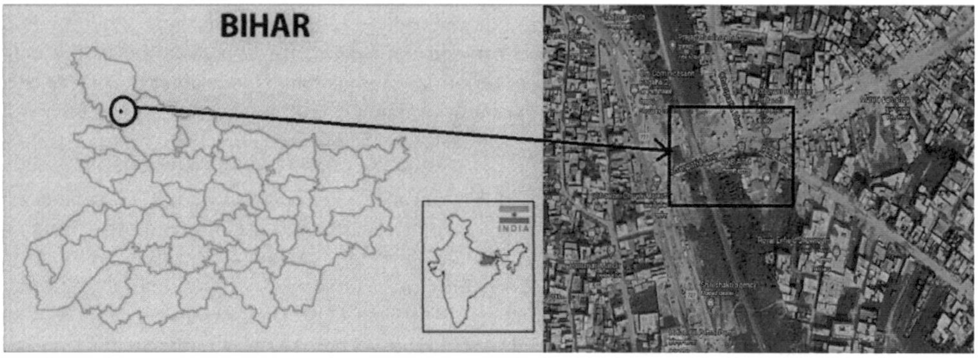

Figure 2. Location of study area.

3 Soil Exploration

Soil exploration is required to ascertain various sub-soil characteristics, stratification data and engineering properties. Figure 3 presents the locations of ten boreholes. The entire exploration program has been divided broadly into two parts, fieldwork and Laboratory Tests. Fieldworks unfold the sub-surface deposit types and their characteristics through the samples obtained by advancing the boreholes of 150 mm diameter by auger and rotary equipment up to 40m depth [12]. The Standard Penetration Test (SPT) value is one of the indicators of strength of soil [13, 14]. Thus, the scope also included conducting SPT [15], collecting disturbed samples at regular intervals for identification and logging purposes, collecting undisturbed tube samples at suitable intervals of 1.5 m or at the change of strata whichever is earlier, and testing these in the laboratory.

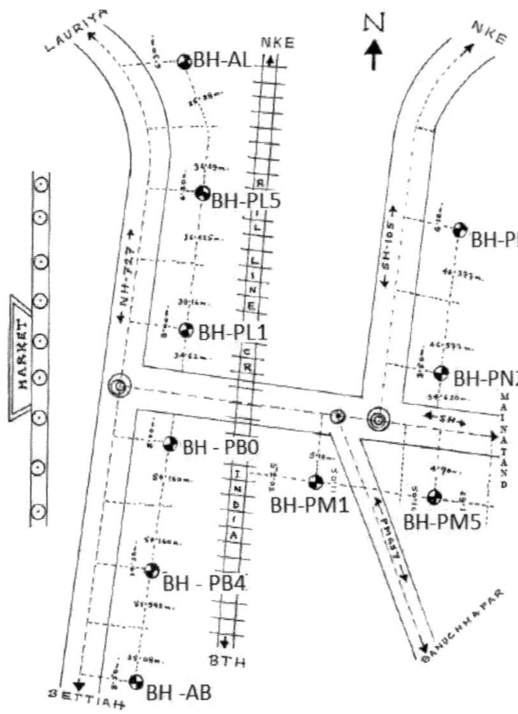

Figure 3. Location of boreholes.

Laboratory tests were performed at NIT Patna laboratory to determine water content, bulk density, specific gravity, and strength parameters (Table 1 and 2). The water table was at an average depth of 1.2 m from the top of the ground level. During investigation it was found that SPT value at PB-0 and AB site (Fig. 4) changes drastically after 20 m length which is also reflected later in the ULLC of pile. Based on the visual classification and investigation results, of field and laboratory tests on the samples recovered, we divided the proposed site into six layers. Overall soil profile can be described as stiff clay to dense sand and geotechnical properties of profiles is presented in Table 1.

Table 1. Major soil strata available at proposed site.

Layers	Thickness (m)	Field N value	IS Classification	Bulk Density (gm /c.c.)
1	0.0 – 1.4	-	CL-ML	1.82
2	1.4 – 5.5	4 – 9	CL, ML, CL-ML	1.83 – 1.86
3	3.65 – 13.45	10 – 18	SM, SM-SP	1.84 – 1.86
4	4.2 – 12.2	15 – 34	SP, SM-SP	1.86 – 1.88
5	6.0 – 13.7	32 – 53	SP, SM-SP	1.87 – 1.89
6	6.6 – 15.25	55 – 100	SP, SM-SP	1.88 – 1.91

Table 2. Average boreholes properties.

Borehole	Natural Moisture Content (%)	Bulk Density (gm/cc)	Specific Gravity (G)	Cohesion (c) (kg/cm^2)		ϕ (degree)	
				CL-ML	SM-SP	CL-ML	SM-SP
PM-1	18.44	1.88	2.65	0.255	0	4.50	32.330
PB-0	18.92	1.88	2.65	0.380	0	4.00	33.125
PB-4	18.72	1.87	2.65	0.290	0	5.00	32.890
AB	18.42	1.88	2.66	0.435	0	3.00	32.780
PM-5	18.56	1.89	2.65	0.293	0	3.67	32.500
PL-1	18.30	1.88	2.65	0.423	0	4.00	32.875
PL-5	18.12	1.90	2.66	0.340	0	4.00	32.750
AL	19.48	1.87	2.65	0.323	0	6.50	33.125
PN-2	18.46	1.90	2.65	0.335	0	4.00	32.330
PN-6	18.60	1.89	2.66	0.370	0	3.50	33.110

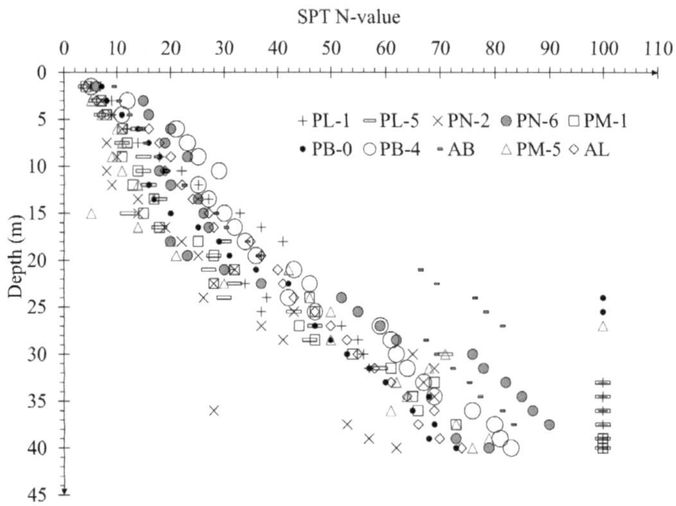

Figure 4. Scatter of SPT N-value.

4 Methodology

To study the behaviour of laterally loaded piles different methodologies such as Subgrade Reaction approach, Ultimate Limit State method, Continuum approach and Finite Element approach have been reported in literature [3]. The present study utilizes the subgrade reaction approach utilized by Indian standard method and M&R method for calculation of ULLC. Further, Results obtained from both the methods were varied against various combination of length, diameter, head condition and concrete grade of pile to understand the respective influence on ULLC.

4.1 Indian Standard (IS) Method (2010)

This method is performed in accordance with Indian Standard [11] & it provide an approximate solution due to the complexity of the problems. The procedure starts with determination of the behaviour of piles considering a flexible member of infinite length or a short rigid pile [5, 11, 16]. Firstly, the Stiffness Factor (T or R) is computed for the specific combination of pile and soil which is further linked to the grade of concrete, dimension of pile and Modulus of Subgrade Reaction whose values are taken from IS 2911 which is listed in Table 3. The site in the present study consists of sand primarily; thus, stiffness factor (T) is calculated (Eqn. 1). As per IS criteria (Table 4), all pile comes under the range of long pile. Thus, the present study encompasses the behaviour of long piles. Virtual Fixity depth indicates a specific profundity below the ground surface where piles are fixed and not meant to yield under lateral load application. Further, the Depth of virtual fixity is used in calculating the lateral deflection of piles and bending moment in the conventional elastic analysis. Further, the cantilever distance between the ground and the point of application of lateral load is calculated. Further, Eqn. 2 and Eqn. 3 has been used to calculate the ULLC for piles with free head and fixed head, respectively.

$$T(m) = \sqrt[5]{\frac{EI}{\eta_h}} \quad (1)$$

$$H(kN) = \frac{3EIy}{\left(e+z_f\right)^3} \times 10^{-3} \quad (2)$$

$$H(kN) = \frac{12EIy}{\left(e+z_f\right)^3} \times 10^{-3} \quad (3)$$

Where, E stands for modulus of elasticity of pile material in kN/m^2, 'I' is moment of inertia of the pile cross-section, in m^4, η_h is modulus of subgrade reaction (MN/m^2), 'z_f' is depth to point of fixity, 'y' is the maximum pile head deflection in mm and 'e' is the cantilever length above ground surface to the point of application of load in meter.

Table 3. Modulus of subgrade reaction for granular soil η_h (kN/m^3) [10].

S. No.	Soil type	SPT N-value (Blows/30 cm)	Range of η_h (kN/m^3)	
			Dry	Submerged
1	Very loose sand	0 – 4	< 0.4	< 0.2
2	Loose sand	4 – 10	0.4 – 2.5	0.2 – 1.4
3	Medium sand	10 – 35	2.5 – 7.5	1.4 – 5.0
4	Dense sand	> 35	7.5 – 20	5.0 – 12.0

Table 4. Criteria for behaviour of pile based on its embedded length [10].

SI. No.	Type of pile behaviour	Relation of embedded length with stiffness factor	
		Linearly increasing	Constant
1	Short (rigid) pile	$L \leq 2T$	$L \leq 2R$
2	Long (elastic) pile	$L \geq 4T$	$L \geq 3.5R$

4.2 Matlock and Reese Method (1961)

Reese and Matlock [5] proposed to use the principle of dimensional analysis to establish the non-dimensional relationships for piles. Further, Matlock and Reese [16] proposed equations for the determination of deflection, slope, shear force, bending moment and soil reaction at any point along the depth of the pile and M&R [7] introduced a non-dimensional Deflection Coefficient for calculating lateral deflection of the pile. For very long piles, it is observed that the total length 'L' loses its significance because the deflection is almost zero for much of the length of the pile [5]. It is convenient to introduce some characteristics length as a substitute. A linear dimension

'*T*' is therefore included in the analysis. Since T in the analysis expresses a relationship between the soil stiffness and the flexural stiffness of the pile material, it is called the relative stiffness factor. For an applied lateral load and moment, the outcome for the deflections of the elastic curve will include the relative stiffness factor and expressed as Eqn. 4. An elastic behaviour can be assumed for minor deflections with respect to the pile dimensions, for which the principle of superposition can be applied. By applying the principle of superposition, the effects of imposed lateral load (H_t) and imposed moment (M_t), if considered separately, give rise to deflection. Deflection at any point x from top of the pile, having flexural rigidity EI can be accessed as per Eqn. 5.

$$y = f_y\left(x, T, L, E_s, H_t, M_t\right) \quad (4)$$

$$y_x = \frac{H_t T^3}{EI} A_y + \frac{M_t T^2}{EI} B_y \quad (5)$$

Where, E_s is elastic modulus of soil, EI is flexural rigidity of pile, A_y and B_y are deflection coefficient given by [7] and its values can be chosen from the table provided by M&R [7] for a given depth coefficient (x/T).

Free Head Pile. When the pile head is free to rotate, the corresponding equations for deflection at ground level is expressed as Eqn. 6. In free head case, pile head is in pinned condition thus moment due to fixity will not develop [Eqn. 7].

$$y_g = 2.435\frac{H_t T^3}{EI} + 1.623\frac{M_t T^2}{EI} \quad (6)$$

$$y_g = 2.435\frac{H_t T^3}{EI} \quad (7)$$

Pile Head Fixed Against Rotation. When Degree of Freedom is zero for a pile at its head then the general equation of deflection is expressed as Eqn. 8 and Moment developed at ground level due to fixity is given by Eqn. 9. Using the value of Deflection coefficient A_y and B_y for zero depth coefficient is 2.435 & 1.623, respectively [7], lateral load can be reported in form of Eqn. 10.

$$y_x = \frac{H_t T^3}{EI} C_y \text{ where, } C_y = A_y + \frac{M_t}{H_t T} B_y \quad (8)$$

$$M_t = M_g = -0.93 H_t T \quad (9)$$

$$H_t = \frac{EI y_g}{0.93 T^3} \quad (10)$$

5 Results and Discussion

ULLC was calculated for all the ten boreholes by using IS code [11] & Matlock & Reese method [7], and considering variation in number of parameters like diameter, length, grade of concrete and pile head condition.

5.1 Comparison between results of IS code and Matlock & Reese method

M&R [7] gives a better representation of field test results [17]. Reese [8] utilized concept of M&R [7] for lateral load calculation and found very less variation when compared with actual field results. Overall, Highest value of ULLC was 1095 kN (M&R), obtained at PB-0, for a M40 grade pile of 1.2m diameter and 25 m length under fixed head condition, and lowest value was 4.3 kN (IS 2911) at borehole PL-5, for the M25 grade pile having 0.4 m diameter and 15m length under free head condition. (Fig. 5). This can be attributed to sudden increment of N-value after 20 m for PB-0. Highest & lowest value of ULLC was obtained corresponding to L/D ratio of 20.83 & 37.5, respectively, for all the borehole sites but PL-1. PL-1 exhibits higher SPT N-value at lesser depth of the pile (15m & 18m) and lower value of that at greater depth of the pile (21 m & 25 m), due to which it shows highest & lowest value of ULLC for L/D ratio 15 & 52.5 respectively. It was observed that the range of ULLC was largest for the calculation with M&R method for fixed head. M&R [7] gives ULLC value higher than IS code method [11] for both head conditions.

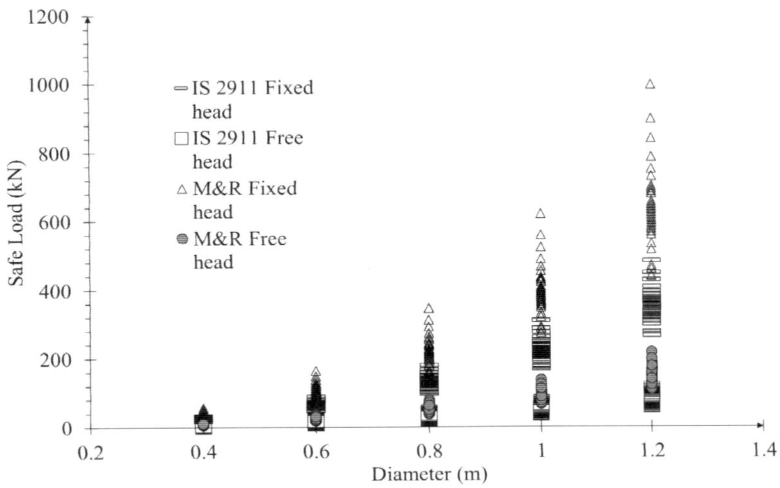

Figure 5. Scatter of ULLC.

5.2 Effect of Diameter

To check the effect of pile diameter on ULLC, the different pile diameters such as 0.40 m, 0.60 m, 0.80 m, 1.0 m and 1.20 m were considered. Fig. 6 and Fig. 7 shows response of pile diameter on ULLC at each borehole using IS code for fixed & free pile head respectively. It is seen that the lateral capacity to carry safe load increases with increase in pile diameter. An approximate increment of 150% & 140% was observed in ULLC for fixed and free head conditioned pile respectively, for 200% increment in diameter. This is attributed to the increase in moment of inertia, stiffness factor, surface area and depth of fixity which relies upon the diameter of the pile. It was observed from the increasing slope of the curves in Fig. 6 & Fig. 7 that the rate of increment of lateral capacity of the pile increases with increase in diameter and a similar response was reported by Salini et al. [18]. Average variation between ULLC calculated by IS 2911 and M&R method increases from 55% to 75% for a diameter change from 0.4m to 1.2 m respectively.

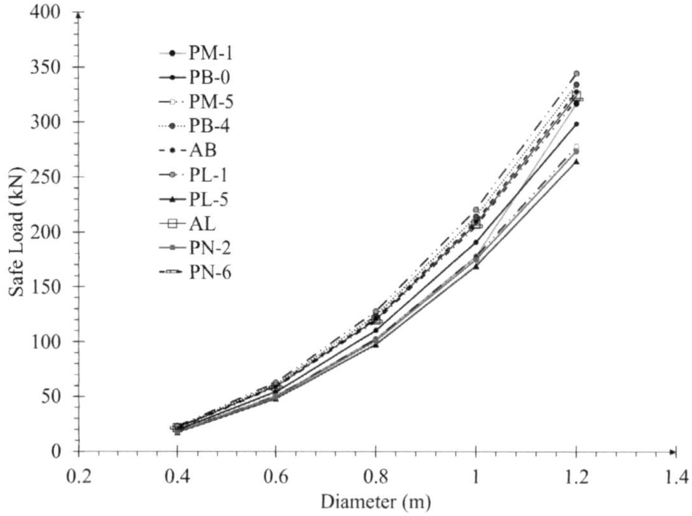

Figure 6. ULLC evaluated as per IS 2911 [10] approach (Fixed head; M25; L = 15 m).

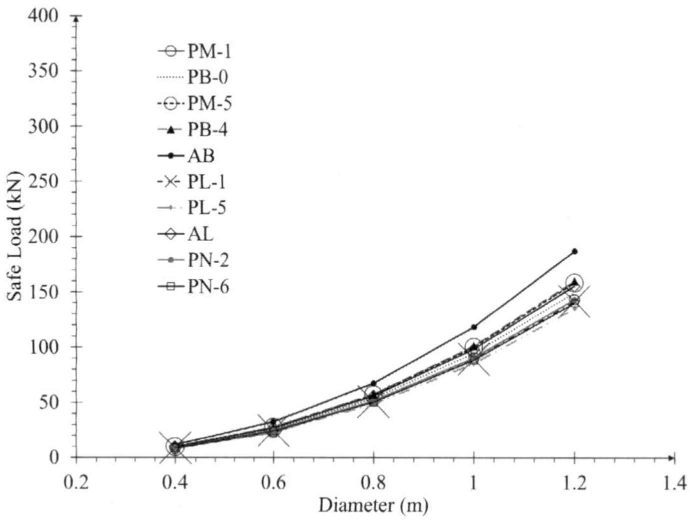

Figure 7. ULLC evaluated as per M&R [7] approach (Free head; M25; L = 21 m).

5.3 Effect of Length of the Pile on ULLC

To examine the impact of length of the pile on ULLC, distinctive pile lengths such as 15 m, 18 m, 21 m and 25 m were considered in present paper and a typical result is presented in Fig. 8. The figure shows non-linear behaviour at lesser depth and becomes relatively flatter at higher depth. This behaviour is associated with the fact that ULLC of pile increases at high rate for initial increment in lengths but later it increases linearly [18]. At lesser depth pile behaves nonlinearly with the depth but as depth increase it starts behaving more linear [8]. Due to the mobilization of passive resistance on the extended length of pile (15 to 25 m), ULLC increases with increase in the pile length.

Changing the length from 15 m to 25 m leads to lowest increment of 4% for the borehole PL-1 and highest increment of 78% for PB-0. Percentage increment in ULLC due to length, varies between (4%-78%) for fixed head condition & between 3.8% to 67% for free head condition at all the borehole sites and this can be attributed to the non-uniform SPT N-value distribution along depth. It is also obtained that percentage change in ULLC due to length increment is higher for M&R approach.

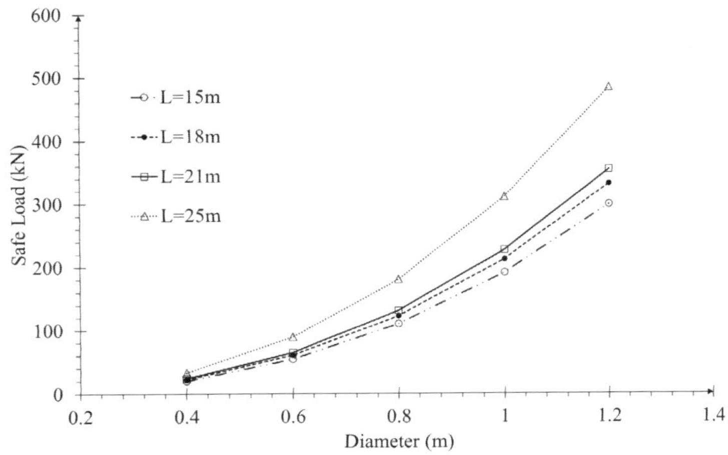

Figure 8. Effect of length on ULLC (PB0; IS2911 [10]; Fixed head; M25).

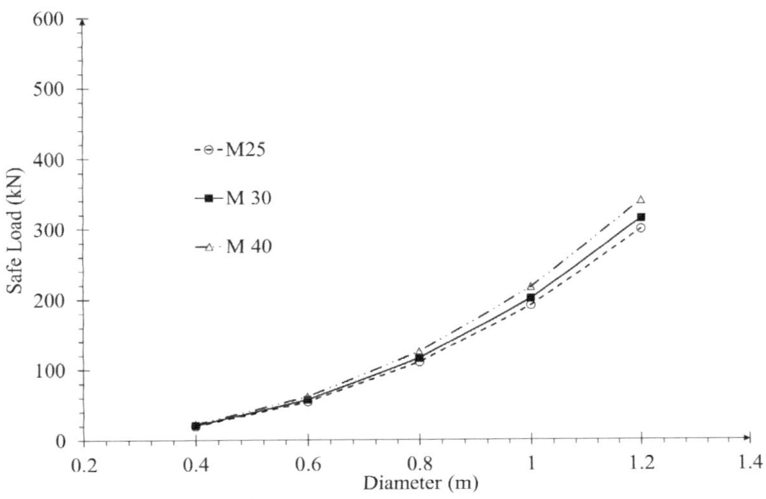

Figure 9. Change in ULLC with concrete grade (PB0, IS 2911 [10], Fixed head, L= 15m).

5.4 Influence of grade of concrete on ULLC

Evaluation of influence of grade of concrete on ULLC is carried out by selecting three different Grade of concrete i.e., M25, M30 and M40. The result shows least increment in ULLC for increase in grade of concrete, when compared with other variable parameters (Fig. 9). When the grade of concrete is increased from M25 to M40, the capacity increased by 13% approximately for both methods and for both head conditions, at each of the borehole site.

5.5 Influence of pile head condition on lateral load capacity

Free head and Fixed head conditions were undertaken to understand the impact of pile head condition on ULLC. It was obtained that pile head condition has significant impact on the lateral load bearing capacity due to fixity of pile which caused an additional moment. When the head condition was changed from free head to fixed head, the ULLC increases by approximately 320% for both the methods. The percentage increment in ULLC due to head condition increases with increase in diameter but more or less remains constant for every borehole site at a particular diameter. The variation is shown in Fig.10 for borehole PB0 and an approximately similar variation pattern was reported for other borehole sites.

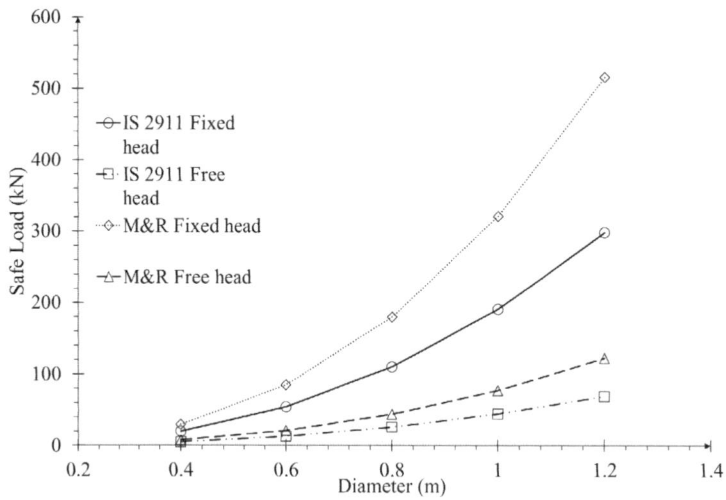

Figure 10. Effect of Head condition on ULLC (PB0; M25; L = 15 m).

6 Conclusions

The ULLC for piles in a cohesionless soil stratum is evaluated for both fixed and free head conditions, using IS Code method [11] and Matlock & Reese [7] method, considering piles of various diameter, lengths and concrete grade. Average variation between both the method (for fixed head condition) increases from 55% to 83% for a diameter change from 0.4 m to 1.2 m respectively. This variation confirms to the results obtained by Jayasree [17]. While, for free head condition this variation ranges between 63% to 84%. From obtained results, it is evident that IS2911: 2010 gives approximately 70% conservative value of ULLC for the case of fixed headed pile and about 75% for free head conditioned pile. Diameter has maximum and grade of concrete has minimum influence on ULLC. Previous studies also find IS code method of lateral load analysis uneconomical [17] and present study is also in the agreement.

• References

1. Winkler, E. (1867). Die lehre von elastizitat und festigkeit (on elasticity and fixity). Dominicus, Prague.
2. McClelland B., Focht J. A. (1958). Soil modulus for laterally loaded piles. Trans. Am. Soc. Civ. Eng. 123(1), 1049–1063.

3. Moussa A. and Christou P. (2018). Advances in analysis and design of deep foundations, sustainable civil infrastructures. Springer International Publishing, DOI 10.1007/978-3- 319-61642-17.
4. IRC 6-Section II (2014): Standard Specifications and Code of Practice for Road Bridges (Loads and Stresses). Indian Roads Congress, New Delhi 110022.
5. Reese L.C. and Matlock H. (1956). Non-dimensional solutions for laterally loaded piles with soil modulus assumed proportional to depth. In 8th Conference Proceedings on Soil Mechanics and Foundation Engineering, University of Texas.1-41.
6. Shah A., Advani P., Patel J. and Soni H. (2017). Estimation of load carrying capacity of bored cast in-situ piles. International Journal of Engineering Technology Science and Research ISSN 2394 – 3386. 4.
7. Matlock H. and Reeses L. C. (1961). Foundation analysis of offshore pile supported structures. In Proceedings of fifth international conference on soil mechanics and foundation engineering, France. 2. 91–97.
8. Reese, Lymon C., Cox, William R., and Francis D. K. (1974). Analysis of Laterally loaded piles in sand. Offshore Technology Conference, Houston, Texas. doi: https://doi.org/10.4043/2080.
9. Davisson, M. T. (1970). Lateral load capacity of piles. Committee on Foundations of Bridges and other Structures. 49th Annual Meeting.
10. Birid, K. C. (2015). Pile head deflection analysis based on theory, software and field tests. 50th Indian Geotechnical Conference. Pune. India.
11. IS 2911-Part 1, Section-2 (2010). Design and construction of pile foundations (Bored cast in-situ concrete piles). Bureau of Indian Standards, New Delhi.
12. IS 1892 (1979). Code of practice for subsurface investigations for foundations (R2002). BIS. New Delhi.
13. Kumar, R., Kapilesh, B. K., and Choudhury D. (2016). Estimation of engineering properties of soils from field SPT using random number generation. Indian National Academy of Engineering.
14. Bowles, J. E. (1968). Foundation analysis and design. 2nd edn. McGraw-Hill Book Company, New York.
15. BIS 2131 (1963). Indian standard method for standard penetration test for soils, Bureau of Indian Standards. New Delhi.
16. Matlock, H and Reese, L. C. (1960). Generalized solutions for laterally loaded piles. Journal of Geotechnical Engineering, 86, 63-91.
17. Jayasree, P. K., Arun, K. V., Oormila, R. and Sreelakshmi, H. (2018). Lateral load capacity of piles: a comparative study between indian standards and theoretical approach. J. Inst. Eng. India Ser. A. 99(3). 587–593.
18. Salini, U. and Girish, M. S. (2009). Lateral load capacity of model piles on cohesionless soil. EJGE. 14.

Impact on the Geotechnical Properties of Soil mixed with Terrazyme

Rishi Ranjan[1] and Rakesh Pratap Singh[2]

[1]Ph.D. Student, National Institute of Technology, Jamshedpur-831014,
rishi43031@gmail.com
[2]Associate Professor, National Institute of Technology, Jamshedpur-831014,
rpsingh.ce@nitjsr.ac.in

Abstract. The financial sustainability and serviceability of any project after standard performance are the most crucial requirements in emerging nations like India. Hence there is a need to find the best possible ways to satisfy the performance as well as economic criteria. The stabilization of soil with bio-enzyme is a revolutionary technique which becoming popular worldwide. Recently there are many bio-enzymes available for soil stabilization such as Fujibeton, renolith, Perma-Zyme, Terrazyme, etc. These enzymes have been proven to be very effective and economical as well as environment friendly in nature. The efficiency of bio-enzyme depends upon the amount of dosage, types of soil and curing period. One such type of bio-enzyme, Terrazyme has been used in the present work. This study shows the effect of Terrazyme on stabilization of soil. Including the improvement in unconfined compressive strength, shear strength and the Atterberg's limits. The enzyme treated soil shows significant improvement in strength parameters of soil and little improvement in liquid and plastic limit of the soil.

Keywords: Strength; Non-traditional soil stabilization; Terrazyme;

1. Introduction

Soil is the main element in this nature. Everyone has a relationship with it in some way. All of life's basic essentials, such as food, clothes, and housing, have been made available by the land. In general, it is challenging to imagine life in this world without soil. Soil is crucial to our lives, and almost every component of construction uses soil in some or other way. However, now soil stabilization is very important for construction stages like dam construction, road construction, and similar works. By adding and combining various materials to the soil, soil stabilization is the procedure for improving the soil's properties to gain long-term stability of soil, and shear strength parameters of soil and therefore also increasing the soil's capacity to bear load transferred by structure. It is typically necessary when the soil beneath the construction site's foundation is unsuitable for supporting the structural load. To increase the geotechnical characteristics of poor soils, such as compressibility, strength, permeability, and durability, stabilizing chemicals are used. This can be done by adding stabilizer ingredients with natural soil to create a homogeneous combination. This technique is a corrective approach that decreases the soil's permeability and compressibility in earthen structures in order to raise the shear strength also helps to minimize the settlement of foundations.

Previous studies have shown that the stabilization of soil with bio-enzyme, requires some curing period to gain full strength (Lacuoture and Gonzalez; 1995) and response of Terrazyme treated soil in case of cohesive soils were observed to have insignificant variation in properties in the earlier stage. Brazetti and Murphy (2000) describe the benefits of adopting bio enzymes as stabilizers to improve the structure and reduce weaknesses in the upper layer surface of the road due to improvement in the CBR value of the sub-base course. Bergmann (2000) concludes that the clay content is necessary for Bio-Enzymes to enhance the soils' properties. It was shown that stability requires at least 2 percent of clay, whereas 10 to 15 percent clay produced excellent results. Shukla et al. (2003 shows that the clay content in soil improves it to vary from low to high strength and the pavement thickness is reduced by 24% to 48%. Sharma (2006) conducted laboratory studies on high plastic clay (CH), low plastic clay (CH), and silt of low plasticity (ML) mixed with Terrazyme and reports that the CH soil had an increase in CBR value with the reduction in saturation moisture content from 40% to 21% after 4 weeks of stabilization. Naagesh and Gandgadhara(2010) presents that the bio-enzyme stabilizer on an expansive soil minimizes the swelling in expansive soil. Li et al. (2011) conducted a comparative study on stabilization of Terrazyme and cement and its results show that the Terrazyme stabilization is cost effective as compared to cement stabilization. Greeshma et al. (2014) shows the behavior of clay near its liquid limit with Terrazyme. Agarwal and Kaur (2014) established the ideal dosage of Terrazyme to find the ideal dosage value. Sen and Singh (2015) gives that the stabilization of black cotton soil with enzyme greatly increases the strength of soil as well as bearing capacity and also the resistance to deformation. Shaka et al. (2016) presents the laboratory testing results and shows little decrease in liquid limit and plasticity index with the increase in dosage of Terrazyme.

This study's primary goal is to find the best cost-effective way to stabilize the locally available soil. There are many stabilizing agents on the market; Terrazyme is one of these that has been used in the study.

2. Material used

The engineering soil's strength and shear properties are required for the good quality of the infrastructure to be developed over the soil. It encourages the nation's social, economic, and cultural development. However, experts discovered that this many soils do not meet the typical requirements for the engineers as it is, and it is required to improve the soil properties in situ.

2.1 Local Soil

The majority of the soil present in Adityapur, Jharkhand is acidic, reddish-yellow, finely grained, very permeable, and has a low water-holding capacity. Some basic properties of Adityapur, (Jharkhand) Soil are given in Table-1.

Table 1 Properties of local available soil.

S.No	Properties	Value
1	Specific Gravity(**G**)	2.67
2	Atterberg's limits	
	Liquid limit (**W_L**)	38.62 %
	Plastic limit (**W_P**)	24.57%
	Plasticity index(**I_P**)	14.05%
3	Max dry density(**MDD**)	18.23 **kN/m^2**
4	Optimum moisture content (**OMC**)	11.82%
5	CBR	
	Un-soaked	3.44 %
6	Uniform confined strength (**UCS**)	71.49 **kN/m^2**

2.2 Terrazyme

A natural, non-toxic liquid enzyme composition called Terrazyme improves the chemical and physical characteristics of soil through closer particle bonding and higher compaction densities, this enzymatic activity boosts the geotechnical properties and increases soil stability. When Terrazyme is introduced into soil, it facilitates the decomposition of the organic molecules and improves the soil particles' potential for wetting and bond properties. It facilitates the ability of soil components to be more easily saturated and compacted. These properties of Terrazyme are helpful for improving soil properties.

Table 2 Properties of Terrazyme used as a stabilizer.

Properties	Value
Boiling Point(°**F**)	212
Specific Gravity(**G**)	1.05
Solubility in water	100%
Appearance	Brown

In this investigation, untreated soil as original is evaluated for the fundamental characteristics and then the Terrazyme treated soil is tested for unconfined compressive strength (UCC), and California bearing ratio (CBR). Fig. 1 shows that the sequence of the test followed in the laboratory.

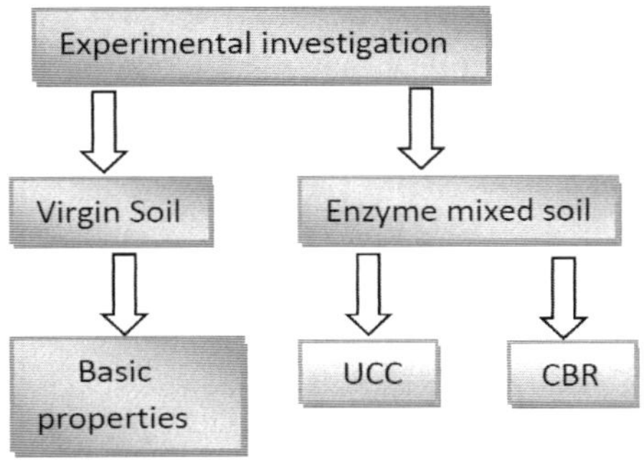

Figure 1. Experiments performed on soil.

3. Results and Discussion

Results of test performed on Terrazyme mixed soil to improve soil characteristics, including the test results of unconfined compressive strength (UCS) and CBR properties of soil with various Terrazyme dosages are presented hereunder.

3.1 Effect of Terrazyme on Unconfined Compressive Strength (UCS)

The force per unit area at which the cylindrical specimen of a cohesive soil falls in compression is known as the unconfined compressive strength (q_u), and 71.49 kN/m^2 is the unconfined compressive strength value of the untreated soil sample. In this work, we mix soil with 0.2, 0.3, and 0.4 milliliters of Terrazyme dosages. The mixture of Terrazyme and soil showed increased unconfined compressive strength as given in the graph below in Fig. 2.

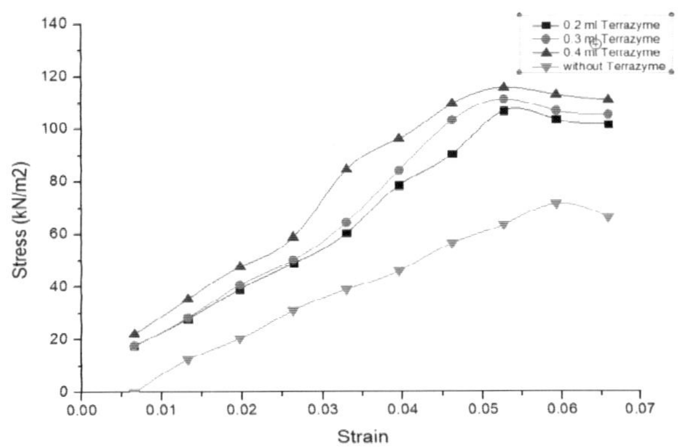

Figure 2. Stress vs. strain of soil under different dosages of Terrazyme.

The maximum value of the unconfined compressive strength (q_u) found in the laboratory of untreated soil (without Terrazyme), and mixed soil with Terrazyme of different dosages as 0.2ml, 0.3ml, 0.4ml are 71.49 kN/m^2, 106.70 kN/m^2, 111.11 kN/m^2, 115.52 kN/m^2 respectively. These results show the improvement of untreated soil with Terrazyme which may be acceptable for constructing super-structures used over treated soil.

3.2 Effect of Terrazyme on California bearing ratio (CBR)

CBR is the ratio of the amount of pressure per unit area needed to penetrate a soil mass with a widely dispersed circular plunger at a strain rate of 1.25 mm/min to the amount of pressure needed to achieve the same penetration in a standard material. The loads corresponding to 2.5 mm penetration and 5 mm penetration are most frequently used to decide the CBR value.

The CBR value for untreated soil is 3.44% at 2.5 mm penetration and 3.18% at 5mm penetration and the CBR value of soil mixed with Terrazyme with different dosages of 0.2ml, 03ml and 0.4ml are

4.2%, 6.3% and 5.6% respectively. These results show that CBR value is improved with treatment of soil with Terrazyme. Fig. 3 shows the results of CBR tests graphically.

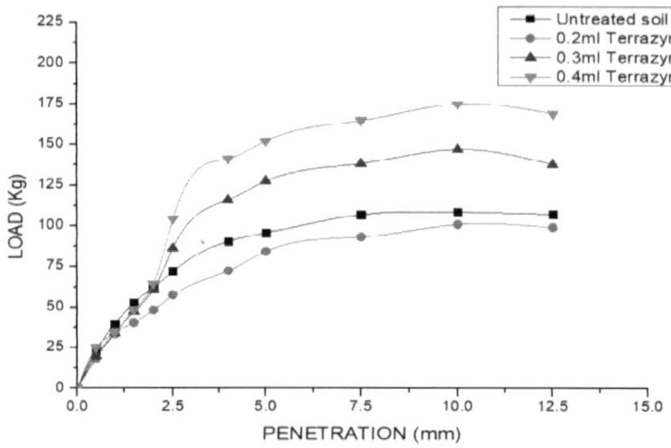

Figure 3. Load vs. penetration curves under different dosages of Terrazyme.

4. Conclusion

The goal of the current study was to investigate the geotechnical characteristics of soil combined with Terrazyme and determine the extent to which these characteristics improve as a result of its addition. The soil was tested with various dosages of the Terrazyme enzyme mixed in unsoaked conditions. Based on an analysis of the experiments and the findings, the following conclusions are drawn.

1. The plasticity index of the soil without Terrazyme is 14.05, and when Terrazyme is applied, the index properties of the soil improve to a good level. As more Terrazyme is added, the index property value also increases.
2. The soil's unconfined compressive strength has improved considerably in Terrazyme-treated soil.
3. Terrazyme-treated soil shows a major improvement in CBR value.

5. References

1. Lacuoture, A., and H. Gonzalez. "*Usage of organic enzymes for the stabilization of natural base soils and sub-bases in Bagota.*" *Pontificia Universidad Jevariana, Faculty of Engineering* (1995).
2. Brazetti, R., & Murphy, S. R. (2000). "*General usage of bio-enzyme stabilizers in road construction*". *Outubro, 15*-19.
3. Bergmann, R (2000) "*Soil stabilizers on universallyaccessible trails*". USDA Forest Service, San DimasTechnology and Development Center.
4. Agarwal, P. & Kaur, S., (2014). : "*Effect of Bio-Enzyme Stabilization on U confined Compressive Strength of Expansive Soil*".International Journal of Research in Engineering and Technology, 03(05), pp.2319–2322.
5. Greeshma,N.E.,Lamanto,T.S.,chandrakaran, S.&Sankar,N.,(2014).: "*Enzyme Stabilization of high Liquid Limit Clay*".Electronic Journal of Geotechnical engineering,19(2014), pp.6990–6994.
6. Khan,T.A.&Taha,M.R.,(2015). :"*Effect of Three Bioenzymes on Compaction ,Consistency Limits , and Strength Characteristics of a Sedimentary Residual Soil*".Advances in Materials Science and Engineering,pp.2-7.
7. Mgangira, M.B., (2009). : "*Evaluation Of The Effects Of Enzyme-Based Liquid Chemical Stabilizers On Subgrade Soils*" .(July), pp.192–199.
8. Ramesh, H.N., Krishnaiah, A.J. & Shet, S.S., (2013). : "*Consolidation Behaviour of Black-Cotton Soil and Mine Tailings Mixtures Treated With Lime*". Proceedings of Indian Geotechnical Conference, pp.1–4.

IOP Conf. Series: Materials Science and Engineering 1273 (2023) 012016 doi:10.1088/1757-899X/1273/1/012016

9. Ramesh, H.N. & sagar, S.R., (2015). : "*Effect of Drying On the Strength Properties of Terazyme Treated Expansive and Non- Expansive Soils*". Proceedings of Indian Geotechnical Conference, pp.20–26.
10. Saini, V. &Vaishnava, P., (2015). : "*Soil stabilization by using terazyme*". International Journal of Advances in Engineering&Technology,8(4), pp.566–573.
11. Sen,J.&Singh,J.P.,(2015).: "*Stabilization of Black Cotton Soil using Bio-Enzyme for a Highway Material*".International Journal of innovative Research in science, engineering and technology, 4(12)pp.12453–12459.
12. Shankar, A.U.R., Rai, H.K. &Mithanthaya, R.I., (2009). : "*Bio-Enzyme stabilised laterite soil as a highway material*".Journal of the Indian Roads Congress . Paper No. 553,(553),pp.143–151.
13. sureka,N.&Gangadhara,S.,(2010). : "*Swelling properties of bioenzyme treated expansive soil*".International Journal of Engineering Studies,2(2), pp.155–159.
14. Taha, M.R. & A. Khan, T., (2013). : '*Recent Experimental Studies in Soil Stabilization with Bio-Enzymes*" – A Review. Ejge, 18, pp.3881–3894.

IOP Conf. Series: Materials Science and Engineering 1273 (2023) 012017 doi:10.1088/1757-899X/1273/1/012017

Numerical Study on the Reduction of Dynamic Lateral Earth Pressure on Retaining Wall using Waste Tyre

[1]Vikash Singh, [2]Samreen Bano, [3]Rakesh Kumar, [4]Vivek Kumar Tripathi, [5*]Anand Verma

[1,3,5]Research Scholar, Department of Civil Engineering, Institute of Engineering and Technology, Lucknow 226021, Uttar Pradesh, India.

[2]Research Scholar, Department of Civil Engineering, Integral University, Lucknow 226026, Uttar Pradesh, India

[4]Assistant professor, Department of Civil Engineering, Institute of Engineering and Technology, Lucknow 226021, Uttar Pradesh, India.

*Corresponding Mail - anandverma3192@gmail.com

Abstract

The performance of retaining under various loading circumstances is dependent on the kind of backfills employed; nonetheless, several lightweight fill materials such as EPS geofoam, geoboard, fly ash, waste tyre (in various forms), and so on are widely utilised as backfill materials. The key benefits of adopting lightweight materials are a lower overall lateral force on the wall and lateral displacement of the retaining wall. Because of the growing number of automobiles, disposing of discarded tyres has become a major issue. Waste tyres are increasingly being utilised in geotechnical applications such as embankment fill, retaining walls, machine foundations, and bridge abutments. According to prior research, shredded tyre inclusion with soil is employed as backfill material for earth-retaining constructions.

The most important aspect of sectional design for retaining walls is the total lateral thrust. Literature indicates that the use of waste tyres in backfill decreases the overall lateral thrust whether used as sole backfill as tyre chips or a mixture of sand tyres or when used as a compressible inclusion between the wall and the backfill.

In the present work, a numerical simulation was conducted using OPTUM G2 (a computational tool based on finite elements) to examine the effect of discarded tyres as backfill material on the total lateral earth pressure for an 8-meter-high wall. Maximum surcharge pressure of 20 kPa is applied to the backfilling.

The use of discarded tyres as a backfill material significantly decreases total lateral earth pressure on the wall by 50-54% compared to walls backfilled with soil, according to the current study.

Keywords: - Waste Tyre, Dynamic earth pressure, Retaining Wall, Numerical Modelling.

I. INTRODUCTION

Bridge abutments, mechanically stabilized walls, retaining walls, braced excavation, and bulkheads were all significant earth retaining structures for geotechnical engineers since they play an essential part in transportation and other infrastructure systems. As a result, it's crucial to assess the stability of these earth-retaining buildings. In an earthquake, retaining walls are vulnerable to failure due to dynamic loading conditions. Walls were caused by an inertia effect and lateral earth pressure during significant ground motion.

The type of backfill material used to build a retaining wall affects its performance, both under static and dynamic conditions: cohesionless granular soil is preferred. Lightweight materials such as discarded tyres (in shreds or pieces), plastic bottles, geofoam, fly ash, and other lightweight materials are now used (Cecich V et al. 1996, Humphery 1992, Ravichandran N. 2014, Tweedie JJ et al. 1998 Graettinger AJ et al. 2005). Embankment filling, retaining walls, equipment foundations, drainage of septic fields, as well as cushion material are all geotechnical applications for waste tyres. They can be utilised in producing car components, the surface of playgrounds, and mixing asphalt pavements. According to earlier studies, Earthwork substance for earth-retaining walls is tyre chips combined with dirt. The two primary factors that are used to design retaining walls under seismic conditions are seismic pressure and permitted displacement. Previous studies have shown that including Waste tyres in the backfill, among the wall as well as the backfill, either as a sand-tyre mixture or even as a compressible material, reduces the thrust of lateral on a wall. The dimension of the section on the wall of refuge is determined by a reduced lateral thrust and displacement, which reduces the overall construction cost.

It should be noted that tyre production in India is steadily expanding these days. Many discarded tyres have accumulated in several countries, posing a threat to the environment (Genan Business & Development A/S 2012). Dumping abandoned tyres has become a worldwide worrying situation for all countries, as the stockpiling of such wastes poses an environmental and health risk (Liu et al. 1998, Clark et al. 1991, Hermann et al. 2001).It releases deadly smoke and liquid oil when it catches fire, harming the environment and putting human lives in danger.

As a result, it is critical to recycle such discarded materials received from various sources as an alternative to conventional materials to achieve sustainable development goals. Waste tyres are utilised on multiple civil engineering procedures in three types, according to ASTM (ASTM 2008)D6270, notably tyre crumbs (less than 1cm), tyre chips (between 1cm and 5cm), and tyre shreds (more than 5cm) (which has a length of more than 5cm).

During dynamic loading, the study demonstrates the outcomes from a numerical simulation to investigate a decrease in total lateral pressure benefits when using waste tyres instead of conventional granular soil. After that, numerical simulations of various qualities of backfill materials like soil and waste tyres were performed, and dynamic motion's maximum ground acceleration; based on these findings, a factor of safety (FOS) on high and low bound circumstances is stated clearly.

II. BACKGROUND

Soil reinforcement, retaining walls, a column of stone, a foundation, a road embankment and a leachate layer are all examples of civil engineering methods (Tweedie JJ et al. 1998, Ravichandran N, Cecich V et al. 1996, and Graettinger AJ 2005, Eldin NN 1992, Huggins L 2014,Bhalla G 2010, Vinot V 2013 and Bosscher J 1997). (Reddy and Krishna2015,Garge and O' Shaughnessy, 2000, Xiao et al. 2012, Reddy et al. 2016a, Reddy and Krishna 2013 and Dammala et al. 2015,) looked at the impact of deteriorating tyres on ground pressure reduction. Some researchers investigate the discarded tire's varied qualities. It has been proven that as the amount of rubber in the soil increases, so does the soil's moisture ratio (Senetakis et al. 2012, Li et al. 2016, Ehsani et al. 2015Nakhaei et al. 2012). Tyre chips in the form of shreds have a specific gravity of 1.02 to 1.24 (Humphery et al. 1993); this is half as little as granular soil (Ahmed 1993, Sheikh et al. 2013,Ghazavi et al. 2011). Hydraulic conductivity ranged from 0.2 to 0.075 m/s (Humphery et al. 1993, Ahmed 1993, Xiao et al. 2012,Gottelandetal. 2008, and Foose et al. 1996).Bressette (1984), Benda (1995), and Ahmed (1993 utilised a triaxial test of shear, whereas Humphery et al. (1993), Yang et al. (2002), Cosgrove (1998) and Gebhardt (1997)used a direct shear test (1998). (1995). An angle of friction is mentioned to be between 150 and 380 degrees with a cohesiveness value of 20 kPa.

The best combined sand-Tire chips ratio mixture in various proportions by various geo-engineering procedures is investigated by Reddy et al. (2015a). The findings revealed that the lightweight material with a 20% optimum ratio is superior in terms of strength and compressibility behaviour. According to Rao and Dutta (2006), the aspect ratio of rubber shreds bridges the gap between shred size and internal friction angle. They are modifying the angle of friction and cohesion in a shredded tyre backfilled in a retaining wall at various amplitudes as well as mean periods on a motion input, Shrestha et al. (2016) investigated under seismic shaking, the maximum bending moment and shear force in the stem of a shredded tyre backfilled in a retaining wall. Different forms of discarded tyres, such as chips, Reddy and Krishna (2016) employed compressible inclusions in earth-retaining walls of refuge to reduce deflection and dynamic pressure of the earth; It determined a dimension of the cross-section in the wall. After conducting the study, Reddy and Krishna (2015b) employed soil mixed with recycled tyre chips in a lightweight material backfilled in retaining walls. An outcome demonstrates a 50% to 60% reduction in lateral earth pressure and horizontal displacement.

III. MATERIALS AND THEIR PROPERTIES

For this numerical analysis, a retaining wall with the required proportions and a soil system for both granular and waste tyre backfill must be modelled. The soil utilised was Mohr-Coulomb Soil (M-C Soil) from OPTUM G2. The retaining wall is represented in grid form using the mesh element. The stiff model is used to classify the behaviour of such meshes. Table 1 shows the chosen value of the M-C soils (Elms and Richard, 1979), waste tyres (S Yang et al., 2002), and backfill walls.

Table 1: Material Properties

Properties	Soil	Waste Tyres
Elasticity (MPa)	207.99	1.29
Unit weight (KN/m^3)	20	5.73
Cohesion (kPa)	0	21.6
Poisson's ratio	0.3	0.28
Frictional angle (degree)	30	11

Concrete retaining walls were designed using concrete, which is often utilised in construction. The body of the retaining wall is made of hard material with a unit weight of 25kN/m^3 in this numerical tool.

IV. NUMERICAL SIMULATION

Figure 1 depicts the numerical grid for modelling a rigid retaining wall. A wall is 4 m broad at the bottom and 1 m wide at the top. The height (H) of the wall is assumed to be 8 metres. At varied peak horizontal ground accelerations, underneath the retaining wall, to soften the creation of the complete shear wedge (plastic zone), the backfill length is 2H (=16 m) (Kh). The backfill was subjected to a constant distributed load of 20kPa (surcharge pressure). A backfill face on the retaining wall and backfill system is designed for moving freely in the X direction. A model's remaining faces have limited horizontal movement.

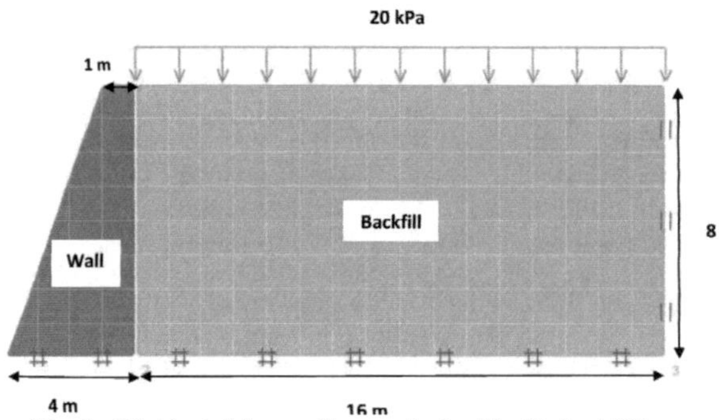

Fig. 1. Rigid retaining wall numerical grid with backfill

Simulation result and discussion

A numerical tool can simulate two-dimensional structural and soil group systems under dynamic loading conditions.

In both soil and waste tyre backfilling situations, Figures 2 and 3 show the contouring for overall bilateral soil pressure through the altitude of a retaining wall which was backfilled with a surcharge at horizontal peak accelerations of the ground of 0.12g I. It is evident from all contours that overall distress grows by up to down in both circumstances; the rise in waste in the tyre, however, is less than that in the earth.

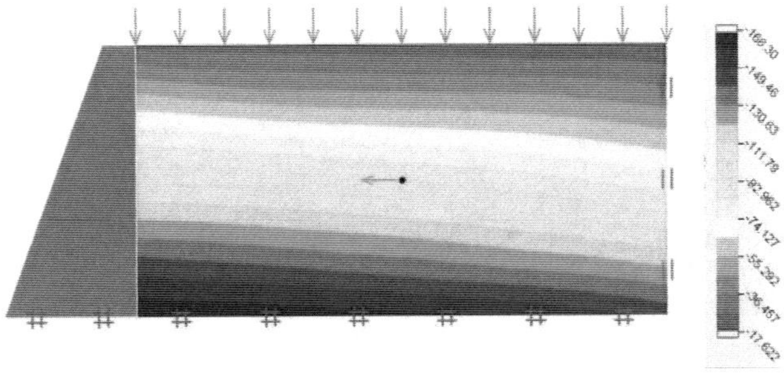

Fig 2: Normal contouring for overall horizontal earth's pressure at a = 0.12g for soil with an 8 meter high wall

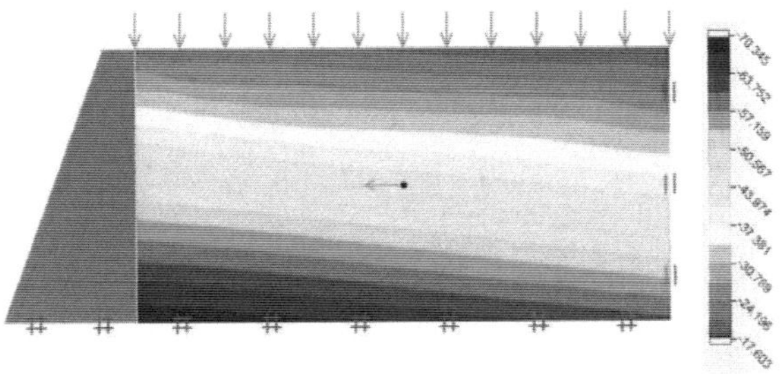

Fig3: For the waste tyre of 8 m high wall, everyday contouring of overall horizontal earth's pressure for k_h = 0.12g

Figures 4 and 5 depict a contour of overall active earth's pressures through the altitude of the retaining wall, which was backfilled to surcharge at a lateral peak with the ground accelerations of 0.24g for both cases whenever the retaining wall has been backfilling with earth and waste tyre, respectively, at horizontal peak ground accelerations of 0.24g. It is evident from all contours that overall stress rises from the top to the bottom for both circumstances, although the number of waste tyres increases was smaller than the increase in the earth.

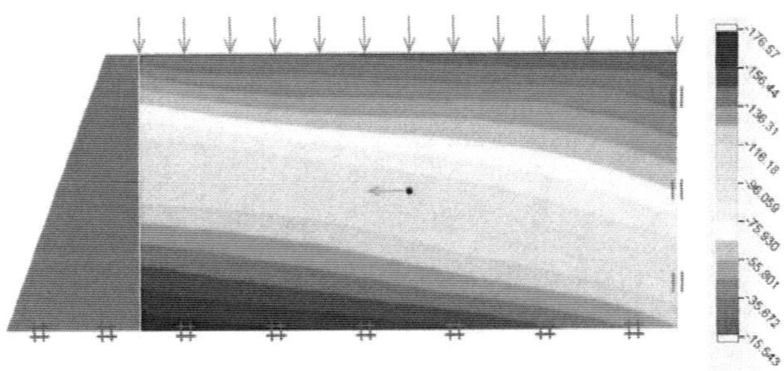

Fig 4: Normal contour for the total horizontal earth's pressure for k_h=0.24g for the soil of an 8m high wall

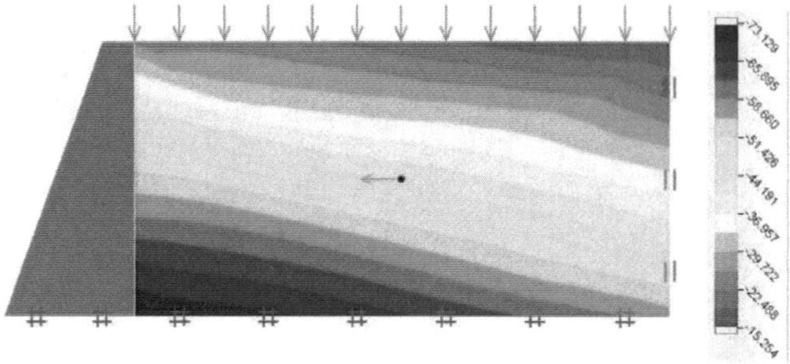

Fig 5: Normal contouring for overall horizontal earth's pressure for k_h=0.24g for a waste tyre of 8m high wall

Figures 6 and 7 depict the contouring of total active earth's pressure mostly through an altitude of a retaining wall which was backfilled to surcharges at lateral peak accelerations of the ground of 0.32g in both cases whenever the retaining wall was being backfilled with earth and waste tyre, respectively, at maximum horizontal ground accelerations of 0.32g.It is obvious from all contours that overall stress rises from the top to the bottom for both circumstances, although the number of waste tyres increases was smaller than the increase in the earth.

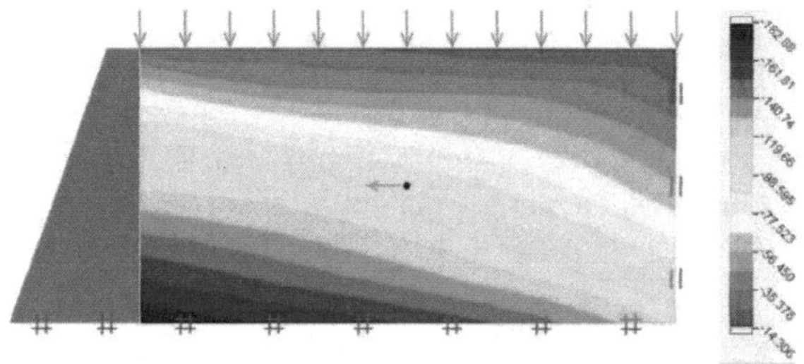

Fig6: Normal contour for the total horizontal earth's pressure for k_h=0.32g for soil of 8m high wall

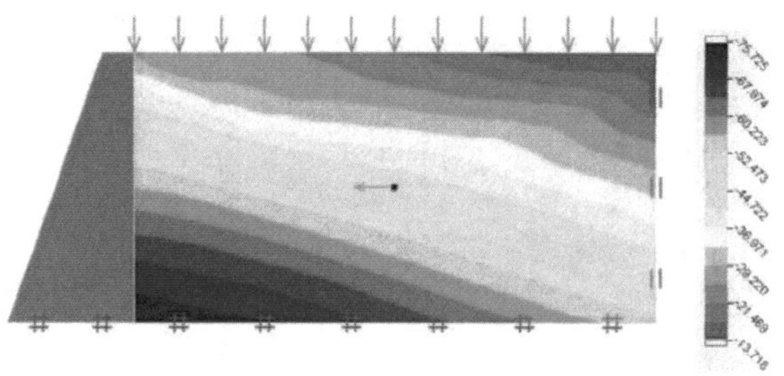

Fig7: Normal contouring for the overall horizontal earth's pressure for k_h=0.32g for the waste tyre of an 8m high wall

Figures 8 and 9 depict the contouring of overall horizontal earth's pressure mostly through an altitude of any retaining wall backfilling to surcharge during maximum horizontal accelerations of 0.42g whenever the retaining wall has been backfilled with soil and waste tyre, respectively, during maximum horizontal ground accelerations of 0.42g.It is obvious from all contours that overall stress rises from top to bottom for both circumstances, although the number of waste tyre increases were smaller than the increase in the earth.

IOP Conf. Series: Materials Science and Engineering 1273 (2023) 012017 doi:10.1088/1757-899X/1273/1/012017

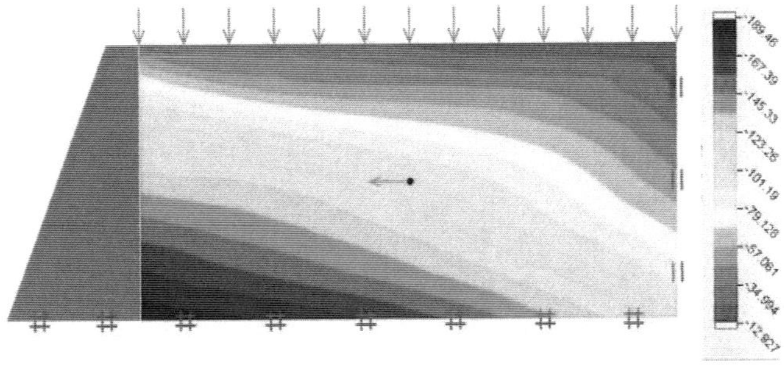

Fig8: Normal contour for the total horizontal earth's pressure for k_h=0.42g for a soil of 8m high wall

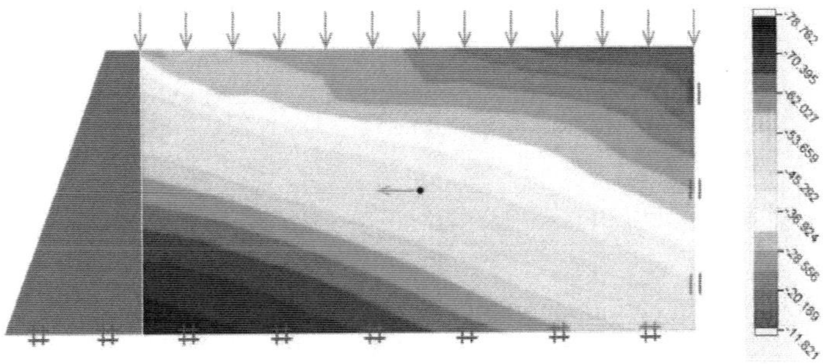

Fig9:Normal contouring for the overall horizontal earth's pressure for k_h=0.42g for a waste tyre of an 8m high wall

Figures 10 and 11 show the contouring of total active earth's pressure mostly through an altitude of a retaining wall which was backfilled with surcharges at maximum horizontal accelerations of the ground of 0.48g, respectively when the retaining wall is backfilled to earth and waste tyre. It is obvious from all contours that overall stress rises from top to bottom for both circumstances, although there was increase in the waste tyre was smaller than the increase in the earth.

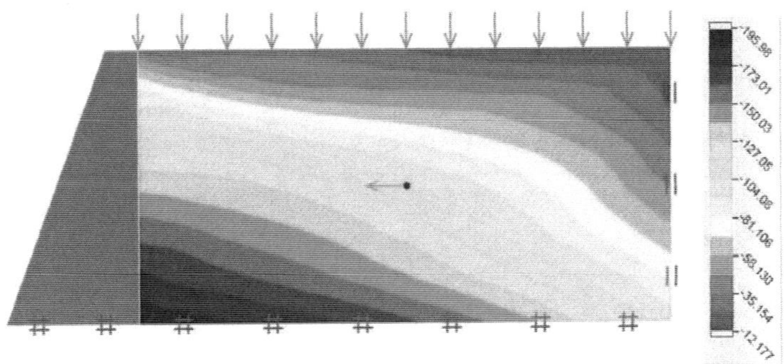

Fig10: Normal contouring for overall horizontal earth's pressure for k_h=0.48g for an earth of an 8m high wall

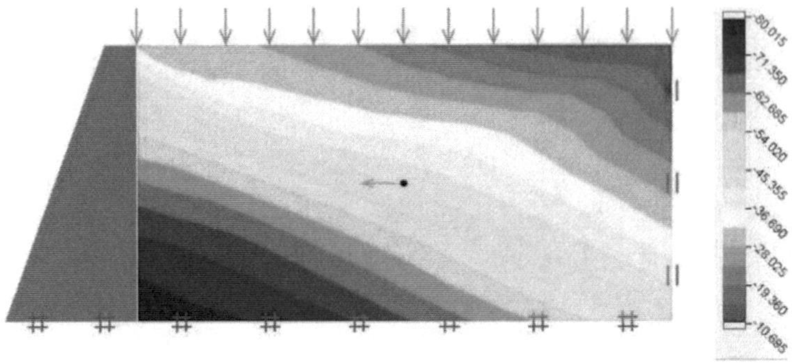

Fig11 Normal contouring for overall horizontal earth's pressure for k_h=0.48g for a waste tyre of an 8m high wall

Table 2 compares a factor of the safety (FOS) of the retaining walls backfilling with the earth and waste tyre for different earthquake excitation levels even at lower and upper bound conditions. FOS on a retaining wall backfill with a waste tyre was considerably greater than FOS on a wall backfill with soil for a given value of K_h, as shown in the table.

Table 2 Identification of factor of safety (FOS) of lower and upper bound conditions on retaining walls

K_h	Walls backfilled with soil		Walls backfilled with waste tyre	
	Bound		Bound	
	Lower	Upper	Lower	Upper
0.12g	9.055	9.227	21.07	22.02
0.24g	4.558	4.620	10.54	10.67
0.32g	3.430	3.475	7.919	8.013
0.42g	2.621	2.660	6.030	6.106
0.48g	2.300	2.336	5.276	5.342

Though the retaining wall was backfilled with dirt, the factor of safety was found to be less than one, even when the maximal horizontal ground acceleration has been used as 0.48g. When the same will be refilled with the waste tyre and simulated for such a highest horizontal ground acceleration of 0.48g, a factor of safety (FOS) was found and it should be greater than five. This determination demonstrates a retaining wall backfilled with waste tyres has a higher safety factor than a wall which was backfilled with earth.

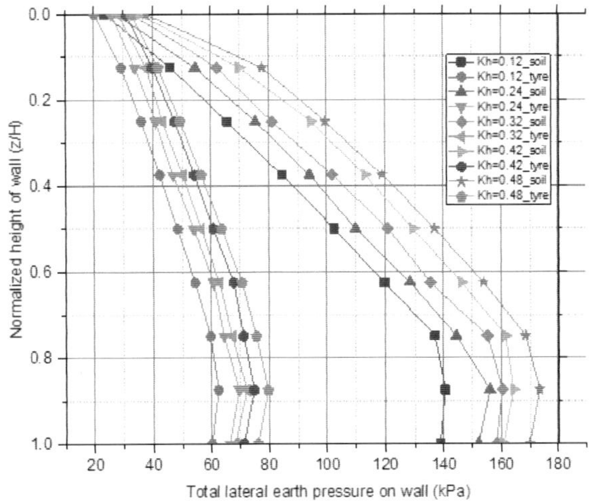

Fig12 Total horizontal earth's pressure acting on a retaining wall

Figure 12 shows a comparative study of the distribution of horizontally acting ground pressure exerted by a retaining wall with soil or a waste tyre. In both the soil backfill and the waste tyre, the total horizontally acting ground pressure on the wall increases from top to bottom. The value of the horizontal earth's pressures may rise from the top to the bottom as the value of peak ground acceleration (Kh) increases in both situations of soil backfill and tyre backfill, however, the decrease in overall thrust was less as the case of the waste tyre as opposed to the sand backfill. By computing each curve's surface area, the total horizontal thrust on the wall was estimated, and the decrease in the total horizontal thrust was calculated. In the case of a waste tyre, the reduction in lateral pressure is just within the limit of 50-54 % to the sand backfill, according to the calculations above.

Conclusion: -

The present study evaluates the effectiveness of rigid retaining walls backfilled with soil and waste tyre under Kh = 0.12g, 0.24g, 0.42g, and 0.48g, in terms of total lateral thrust on an 8m high retaining wall. In all cases walls backfilled with waste tyres are found to be safer than the wall backfilled with soil. From the comparison of the distribution of lateral earth pressure on a wall backfilled with soil and waste tyre, it is noted that for a given value of Kh, lateral earth pressure on the wall reduces substantially when the waste tyre is used instead of soil behind the wall. Furthermore, it is worth mentioning that the reduction in total lateral thrust is observed in the range of 50-54% for an 8 m high retaining wall with a different value of peak horizontal ground acceleration.

Conflicts of Interest: -
The authors declare that there are no conflicts of interest regarding the publication of this paper.

Acknowledgement: -
The authors would like to thank (Optum CE) which was Optum Computational Engineering for making an academic resource available for a free licence for the OptumG2 software, which was used to conduct this research.

References: -
[1] Ahmed, I. (1993). Laboratory study on properties of rubber soils. PhD thesis, School of Civil Engineering, Purdue University., West Lafayette

[2] Benda, C. C. (1995). Engineering properties of scrap tires used in geotechnical applications. Rep. No. 95-1, Vermont Agency of Transportation, Montpelier, VT
[3] Bhalla G, Kumar A, Bansal A (2010) Performance of scrap tire shreds as a potential leachate collection medium. J Geotech Geo Eng 28(5):661–669.
[4] Bosscher J, Edill TB, Kuraoka S (1997) Design of highway embankments using tire chips. J Geotech Geoenviron Eng 123(4):297–304.
[5] Bressette, T. (1984). Used tire material as an alternative permeable aggregate. Rep. No.FHWA/CA/TL-84/07. Office of Transportation Laboratory, California Department of Transportation, Sacramento, CA
[6] Cecich V, Gonzales L, Hoisaeter A, Williams J, Reddy K (1996) Use of shredded tires as lightweight backfill material for retaining structures. Waste Manag Res 14:433–451.
[7] Clark, C., Meardon, K., and Russell, D. (1991) —Burning Tires for Fuel and Tire Pyrolysis‖, Report by Pacific Environmental Services for the U.S. Environmental Protection Agency.
[8] Cosgrove, T. A. (1995). "Interface strength between tire chips and geomembrane for use as a drainage layer in a landfill cover."Proc., Geosynthetics'95, Vol. 3, Industrial Fabrics Association International, MN, 1157–1168.
[9] Dammala P, Reddy SB and Krishna AM (2015) Experimental investigation of the applicability of sand tire chip mixtures as retaining wall backfill. In Proceedings of the International Foundations Congress and Equipment Expo 2015 (IFCEE 2015), San Antonio, TX, USA (Iskander M, Suleiman MT, Anderson JB and Laefer DF (eds)). American Society of Civil Engineers (ASCE), Reston, VA, USA, Geotechnical Special Publication 256, pp. 1420–1429.
[10] Ehsani, M., Shariatmadari, N., and Mirhosseini, S. M. (2015). "Shear modulus and damping ratio of sand-granulated rubber mixtures." J. Cent. South Univ., 22(8), 3159–3167.
[11] Eldin NN, Senouci AB (1992) Use of scrap tires in road construction. J ConstrEngManag (ASCE) 118(3):561–576
[12] Foose, G. J., Benson, C. H., and Bosscher, P. J. (1996). Sand reinforced with shredded waste tires. J. Geotech. Eng., 10.1061/(ASCE)0733- 9410(1996)122:9(760), 760–767.
[13] Garge, V.K. & O'Shaughnessy, V. (2006). Tyre Reinforced Earthfill. Part 1: Construction of a Test Fill, Performance and Retaining wall design. Canadian Geotechnical Journal 37:75-96.
[14] Gebhardt, M. A. (1997). Shear strength of shredded tires as applied to the design and construction of shredded tire stream crossings. M.S. thesis, Iowa State Univ., Ames, IA.
[15] Genan Business & Development A/S. (2012). Scrap tires." ⟨http://www. genan.eu/tyres-2.aspx⟩ (Apr. 18, 2013).
[16] Ghazavi, M., Ghaffari, J., and Farshadfar, A. (2011). Experimental determination of waste tire chip-sand-geogrid interface parameters using large direct shear tests. 5th Symposium.on Advances in Science and Technology, Khavaran Higher Education Institute, Mashhad, Iran.k
[17] Gotteland, Ph., Lambert, S., and Salot, Ch. (2008). Investigating the strength characteristics of tire chips – sand mixtures for geo-cellular structure engineering." Scrap tire derived geo-materials—Opportunities and challenges, Taylor & Francis Group, London.
[18] Gosh, S. (2008). Pseudo-static analysis of rigid retaining wall for dynamic active earth pressure. The 12th international conference of "International Association for Computer Methods and Advances in Geomechanics", October 1-6, 2008, Goa, India
[19] Graettinger AJ, Johnson PW, Sunkari P, Duke MC, Effinger J (2005) Recycling of plastic bottles for use as a lightweight geotechnical material. Manag Environ Qual 16(6):658–669
[20] Huggins E, Ravichandran N (2011) Numerical study on the dynamic behavior of retaining walls backfilled with shredded tires. In: ASCE proceedings of GeoRisk 2011, June 26–28, 2011, Atlanta, Georgia d 2011000, Reston.
[21] Humphrey DN, Sandford TC, Cribbs MM, Gharegrat H, Manion WP (1992) Tire chips as lightweight backfill for retaining walls— phase I. Dept. of Civil Engineering, University of Maine, Orono.
[22] Liu, H., Mead, J., and Stacer, R. (1998). Chelsea —CenterFor Recycling and Economic Development Environmental Impacts of recycling rubber In Light Fill

Applications‖, Summary & Evaluation Of the Existing Literature University of Massachusetts.
[23] Nakhaei, A., Marandi, S. M., Kermani, S. S., and Bagheripour, M. H. (2012). "Dynamic properties of granular soils mixed with granulated rubber."SoilDyn. Earthquake Eng., 43(4), 124–132.
[24] Rao GV and DuttaRK(2006) Compressibility and strength behavior of sand–tire chips mixtures.Geotechnical and Geological Engineering24(3): 711–724
[25] Ravichandran N, Huggins L (2014). Applicability of shredded tire chips as a lightweight retaining wall backfill in seismic regions. In: Proceedings of geo-congress (GSP 234), ASCE, Atlanta
[26] Reddy SB and Krishna AM(2013) Numerical simulations of earth-retaining structures using EPS geofoam inclusions. Proceedings of Indian Geotechnical Conference.IGS Roorkee Chapter, Roorkee, India.
[27] Reddy SB, Kumar DP, Krishna AM (2015) Evaluation of optimum mixing ratio of sand–tire chips mixture for geo-engineering applications. J Mater Civ Eng.
[28] Reddy SB, Kumar DP and Krishna AM(2016a) Evaluation of the optimum mixing ratio of sand–tire chips mixture for geo-engineering applications. Journal of Materials in Civil Engineering, ASCE28(2), paper no. 1335, http://dx.doi.org/10.1061/(ASCE)MT.1943-5533.0001335.
[29] Richard, R.J. and Elms, D. (1979). Seismic behavior of gravity retaining wall.Journal of the Geotechnical Engineering.Division, ASCE, Vol. 105, No. GT4, 449-464.
[30] Senetakis, K., Anastasiadis, A., and Pitilakis, K. (2012)."Dynamic properties of dry sand/rubber (SRM) and gravel/rubber (GRM) mixtures in a wide range of shearing strain amplitudes."SoilDyn. Earthquake Eng., 33(1), 38–53.
[31] Neaz Sheikh, M., J. S. Vinod, M. S. Mashiri, and Hing-Ho Tsang."Shear and compressibility behavior of sand–tire crumb mixtures." Journal of Materials in Civil Engineering 25, no. 10 (2013): 1366-1374.
[32] Shrestha, S. et al. (2016). Design and Analysis of retaining wall backfilled with a shredded tire and subjected to earthquake shaking. Soil Dyn. And Earthquake Engineering 90 (2016), 222-239.
[33] Tweedie JJ, Humphrey DN, Sandford TC (1998) Tire shreds as retaining wall backfill, active conditions. J GeotechGeoenvironEng (ASCE) 124(11):1061–10703. Cecich V, Gonzales L, Hoisaeter.
[34] Vinot V, Singh B (2013) Shredded Tyre-Sand as fill material for embankment applications. J. Environ. Res. Develop. 7(4A): 1622–1627.
[35] Xiao, M., Bowen, J., Graham, M., and Larralde, J. (2012). Comparison of seismic responses of synthetically reinforced walls with tire-derived aggregates and granular backfills. J. Mater. Civ. Eng., 10.1061/ (ASCE)MT.1943-5533.0000514, 1368–1377.
[36] Yang, S., Lohnes, R. A., and Kjartanson, B. H. (2002). Mechanical properties of shredded tires.Geotech. Test. J., 25(1), 44–52.
[37] Zornberg, J. G., Viratjandr, C., and Cabral, A. R. (2004)."The behavior of tire shred-sand mixtures."Can. Geotech. J., 41(2), 227–241.

Strength and stiffness prediction models of expansive clays blended with sawdust ash

Shaik Subhan Alisha[1], T V Nagaraju[2*], Pilla Sita Rama Murty[3], VVS Sarma[1], and Mantena Sireesha[4]

[1] Department of Civil Engineering, Vishnu Institute of Technology, Bhimavaram [2]Department of Civil Engineering, S.R.K.R Engineering College, Bhimavaram, India [3]Department of Information Technology, Vishnu Institute of Technology, Bhimavaram[4]Department of Geo-engineering, Andhra University, Visakhapatnam
*varshith.varma@gmail.com

Abstract. Strength and stiffness characteristics are major concern for selecting any geomaterial. However, laboratory testing of these characteristics is time associated, laborious, and high cost. So, there is a need of intelligence tools to estimate the strength and stiffness of geomaterial. The impact of sawdust ash on the stiffness and strength properties of combined expanding clays is discussed in this research. The combined expansive clays underwent tests for California bearing ratio (CBR), unconfined compressive strength (UCS), optimal moisture content, maximum dry density, plasticity characteristics (liquid limit and plastic limit), and differential free swell (DFSI). According to test results, adding more sawdust to the blended clays improves their performance. This study also investigates the artificial neural network (ANN) model that considers six input variables to forecast the CBR and UCS of blended clays. The findings demonstrate that the ANN model performs more accurately for the CBR and UCS models. This clever method may help manage the under- or overestimation of additive dosage and reduce project costs.

1. Introduction

Since the chemical amelioration technique successfully altered the properties of expansive clays, this technique has been extensively used to counteract swelling and shrinkage of clays. In recent years, many industrial by-products have been gaining importance because the traditional additives cement and lime are energy intensive and emit CO_2 [1, 2]. Industrial by-products such as fly ash, pond ash, rice husk ash, bagasse ash, corn cob ash, cotton hull ash, ground granulated blast furnace slag, and silica fume are successful in controlling heave and improving strength [3, 4, 5]. Chemically altered soils have often been used for pavement subgrade material, embankment film material, and foundation material for lightweight structures [3, 4]. In most of the rural roads in India, altering the properties of locally available materials is desirable to fulfill both strength and economic requirements.

In recent years, large construction projects have accelerated industrial byproducts like fly ash pulverized granulated blast furnace slag and silica fume. Additionally, after 2020, the price of these commodities has grown. Therefore, due to their pozzolanic character and amorphous morphologies, agricultural byproducts such as rice husk ash, bagasse ash, and maize cob ash were gaining relevance [6, 7, 8]. Pozzolanic additives blended with clays allow total exchangeable forms. When the total exchangeable ions attract the cations on the surface of the expansive clays reduces the surface forces. This causes the agglomeration of particles, breaking the double diffusion layer and releasing adsorbed pore water. The

breaking of the double diffusion layer reduces soils' plasticity behavior and swelling behavior [8, 9]. Their many studies reported the utilization of chemical additives and pozzolanic additives to reduce the soils' swelling behavior and improve shear strength [3, 4, 10].
There are few references to using sawdust ash in soils. An experimental examination on black cotton soil stabilized with sawdust ash was published by Onyelowe et al. in 2022. According to the test results, adding sawdust ash to black cotton soil can significantly reduce fracture width, volumetric shrinkage, and linear shrinkage [11]. On the other hand, a limited number of studies indicate that sawdust was effectively used as a landfill liner to hold back heavy metals. The main characteristics of sawdust were its 52% and 31% concentrations of lignin and cellulose, respectively. The landfill liner's stiffness and durability might be enhanced as a result. Additionally, sawdust has the benefit of having a high capacity for heavy metal adsorption. The focus of this research is on how saw dust ash affects the strength behavior of blended clays. Additionally, the use of artificial neural networks to forecast the stiffness and strength properties of sawdust in blended clays, including the California bearing ratio (CBR) and unconfined compressive strength (UCS) (ANN).
Artificial intelligence's use in building materials has grown in relevance recently since it eliminates time-consuming, complex trials manage overestimating additive doses, and lowers project expenses [12]. Using metaheuristic algorithms, it effectively forecasted CBR and UCS values of soils and blended soils. Nagaraju et al. (2020) presented a brand-new metaheuristic swarm-assisted approach to forecast the CBR values of the subgrade soil for a highway project [13]. Alkali-activated blended expanding clays were studied using a swarm-aided algorithm by Nagaraju et al. in 2021. According to evidence from prediction models, PSO models might predict complicated variables with small data sets and higher convergence [14]. To forecast the UCS of chemically treated soils, Tabarsa et al. (2021) used five intelligent techniques, including ANN, random forest (RF), genetic algorithm (GA), and support vector machine (SVM) models. The findings demonstrate that ANN, SVM, and GP models perform better than the RF model. Other writers have attempted to forecast the UCS and CBR of natural and chemically changed soils [15]. The use of sawdust ash is somewhat restricted [9, 16], in contrast to the widespread usage of industrial by-products as a chemical addition to modifying the properties of expanding clays. The impact of sawdust ash on the UCS and CBR values of the combined expanding clays is discussed in this work. This study also investigates the use of the intelligent ANN method to estimate the UCS and CBR values of expanding clays that have been combined with sawdust and ash.

2. Materials and methods

In this work, expansive clay and clay mixed with sawdust ash were the subjects of an experimental examination. In Bhimavaram, India, clay was gathered from a road project at a depth of 0.5 meters. The expansive clay's geotechnical characteristics are displayed in Table 1. In Avapadu, Andhra Pradesh, India, the furfural oil plant produces a by-product from the boiler: sawdust ash. The liquid limit (LL), plastic limit (PL), differential free swell index (DFSI), optimal moisture content (OMC), maximum dry density (MDD), CBR, and UCS of the blended clays with varied sawdust ash (SD) concentration were all determined by trials in this inquiry. The expanding clay and sawdust ash micrographs from scanning electron microscopy (SEM) are displayed in Figure 1.

Table 1. Geotechnical properties of expansive clay

Parameter	Liquid limit, %	Plastic limit, %	Differential free swell index, %	Optimum moisture content, %	Maximum dry density, g/cc	CBR, %	UCS, kPa
Value	82	40	60	24.2	1.68	2.86	290

Figure 1. SEM micrographs a) expansive clay and b) sawdust ash

CBR and UCS are important tests to assess the strength behavior of soils. Moreover, CBR can only help understand the stiffness of the subgrade of pavements, while UCS can be used to design and monitor the quality of engineering projects. UCS and CBR values were determined according to ASTM D5102-09 and ASTM D1883-16, respectively.

To generate prediction models for CBR and UCS, ANN was used. Generally, any supervised machine learning approach requires a dataset of input variables to perform predictions [17]. The dataset includes six input variables, SD, LL, PL, DFSI, OMC, and MDD used to predict UCS and CBR. An artificial neural network (ANN) is a successful tool for performing regression and classification of complex problems [18, 19]. Moreover, this technique is data-driven and processed with neurons and hidden layers. In this proposed ANN model, six input layers, one hidden layer, and two output layers were considered. Table 2 shows the weights matrix representation of the proposed model.

Table 2. ANN model weights matrix

Predictor		Predicted				
		Hidden Layer 1			Output Layer	
		H(1:1)	H(1:2)	H(1:3)	CBR	UCS
Input Layer	(Bias)	-.158	-.424	1.638		
	SD	-.274	-.102	.720		
	LL	-.047	.045	.031		
	PL	.196	-.076	-.130		
	DFSI	.134	.243	1.320		
	MDD	-.311	.159	.291		
	OMC	.138	-.133	-.025		
Hidden Layer	(Bias)				1.429	-.361
	H(1:1)				-.491	-1.936
	H(1:2)				-1.399	-.737
	H(1:3)				-2.349	-.145

In this process, the functions of CBR and UCS represented as equations (1) and (2), respectively.

$$CBR = f(SD, LL, PL, DFSI, OMC, MDD) \quad (1)$$
$$UCS = f(SD, LL, PL, DFSI, OMC, MDD) \quad (2)$$

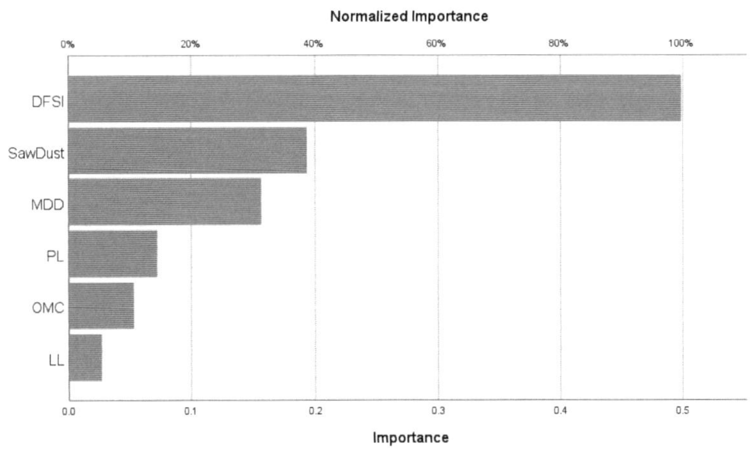

Figure 2. Sensitivity of the input variables

Figure 2 shows the sensitive nature of the input variables for predicting CBR and UCS of the blended expansive clays. From the sensitive analysis, DFSI exhibits higher normalized importance than the other input variables considered. The DFSI values obtained in this study is having higher than 60%. As such, higher DFSI values indicate an inherent relationship between strength, particularly, expansive clays having rich in montmorillonite mineral.

3. Discussion of test results

3.1. Effect of sawdust ash on the CBR and UCS of blended clays

The effect of sawdust ash content on the characteristics of expanding clays is shown in Table 3. In this study, the dry weight of the expansive clays was changed to vary the sawdust ash by 2%, 4%, 6%, and 8%. With increasing sawdust ash concentration, the indices of expanding clay blended with plastic limitations significantly improve. This is because the amorphous sawdust ash tends to push away soil particles and water that has been absorbed.

Table 3. Effect of sawdust ash on the index and engineering properties of clays

Property	Sawdust ash (%)				
	0	2	4	6	8
Differential free swell index (%)	112	96	80	68	44
Liquid limit (%)	84	78	76	72.8	68.5
Plastic limit (%)	22	22.6	23.5	24.8	26.2
Optimum moisture content (%)	24.2	22.6	20.8	19.7	22.6
Maximum dry density (g/cc)	1.68	1.76	1.81	2.04	1.92
California bearing ratio (/%)	3.86	4.62	5.45	8.43	10.15
Unconfined compressive strength (kPa)	290	316	410	495	502

Characteristics of compaction with more sawdust added to the mixed clays, OMC and MDD improved. The addition of sawdust improved OMC values, whereas the addition of sawdust ash improved MDD values by up to 6%. This is because too much sawdust ash can make a mix stiffer, but too much amorphous ash makes blends have a poor matrix, which lowers the MDD value for 8% sawdust ash concentration.

Increased sawdust ash content in the blended clays significantly increases UCS and CBR values. When sawdust ash concentration increased from 0% to 8%, UCS values revealed an increase of 42%. This is brought about by the cementation of blends and amorphous forms, which thin out the double diffusion layer and increase shear resistance. When sawdust ash concentration rose from 0% to 8%, CBR values climbed by 61%. This is brought about by an increase in the blend's rigidity and a decrease in its flexibility. Additionally, 8% sawdust ash mixed with expanding clays might be appropriate for the subgrades of highways.

3.2. Capability of proposed ANN model and performance

The prior section addressed the ANN model's choice of variables and processes. In order to learn the data and establish a connection between the input and output variables, the test data was trained. In order to achieve a predefined mean square error range, training and learning will continue until the neurons recognize and comprehend the data.

Six input data variables were trained for UCS and CBR prediction models until the low mean squared error was attained. Neurons in the hidden layer were acquired during the training of the data. Additionally, during the training phase, the weights and biases of the layers were changed. Regression coefficients for the UCS and CBR models are 0.99 and 0.98, respectively. The regression data show which model best fits the expected and actual values of the UCS and CBR situations. For the ANN prediction model, the relation between actual and predicted values represents linear function and falls on the same line. Moreover, from the prediction point of view, the linear regression indicates the best performance (vide Figures 3 and 4).

A peer glance at the past studies on applying ANN models to predict CBR and UCS. Depending up on the type of the soil and additive added in the blend, the regression performance varies from 0.91 to 0.99 [20, 21]. However, there are no explicit mathematical equations from the ANN model; it can have precession prediction.

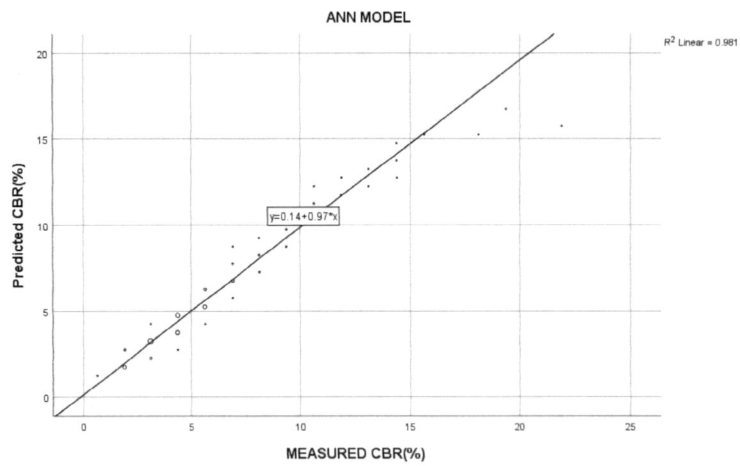

Figure 3. Prediction model of CBR

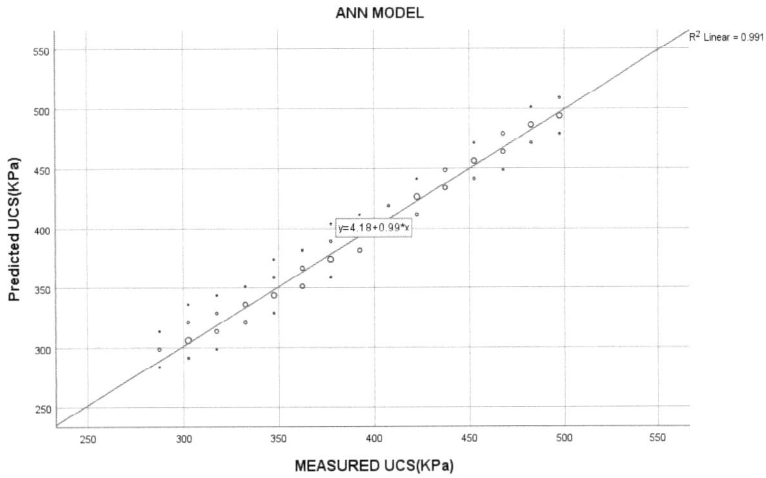

Figure 4. Prediction model of UCS

4. Conclusions

From the current investigation, the following results were drawn:

Sawdust ash performs better than other pozzolanic additions and is an environmentally friendly building resource. Industrial by-products were also used in numerous projects due to the recent introduction of new greenhouse gas taxes on cement supplies, and their cost increased due to increasing demand.

Clays with added sawdust ash have significantly improved plasticity and compaction properties.

As the sawdust ash content of clays increased, so did the UCS and CBR values. This is because more cellulose in the sawdust causes increased Si ions in the sawdust ash. Additionally, these Si ions in the blends permitted cation exchange between clay particles, which contributed to cementitious compounds (C-S-H) creation.

With regression values of 0.99 and 0.98, the ANN prediction models of CBR and UCS performed well. DFSI displays 100% normalized relevance to the relative importance of input factors, which was significant in predicting UCS and CBR values.

5. References

[1] Hills, T. P., Sceats, M., Rennie, D., & Fennell, P 2017 LEILAC: Low-cost CO2 capture for the cement and lime industries. *Energy Procedia*, **114**, 6166-6170.

[2] Bosoaga, A., Masek, O., & Oakey, J. E 2009 CO2 capture technologies for cement industry. *Energy procedia*, **1**(1), 133-140.

[3] Vamsi Nagaraju, T., & Satyanarayana, P. V. V. 2019 Geotechnical aspects of various constructions along the canal embankment using rice husk ash as stabilizer. In *Ground Improvement Techniques and Geosynthetics* (pp. 143-150). Springer, Singapore

[4] Phanikumar, B. R., & Nagaraju, T. V. (2018). Engineering behaviour of expansive clays blended with cement and GGBS. *Proceedings of the Institution of Civil Engineers-Ground Improvement*, *171*(3), 167-173.

[5] Yarbaşı, N., Kalkan, E., & Akbulut, S. 2007 Modification of the geotechnical properties, as influenced by freeze–thaw, of granular soils with waste additives. *Cold regions science and technology*, **48**(1), 44-54.

[6] Raja, K., Venkatachalam, S., Vishnuvardhan, K., Krishnan, R. S. R., Selvan, V. T., & Vetriselvan, N 2022 A review on soil stabilization using rice husk ash and lime sludge. *Materials Today: Proceedings*.

[7] Singh, S 2021 Experimental investigation of corn cob ash on silty clay stabilized with calcium carbide. *Materials Today: Proceedings*, **37**, 3658-3660.

[8] Yadav, A. K., Gaurav, K., Kishor, R., & Suman, S. K 2017 Stabilization of alluvial soil for subgrade using rice husk ash, sugarcane bagasse ash and cow dung ash for rural roads. *International Journal of Pavement Research and Technology*, **10**(3), 254-261.

[9] Sharma, A. K., & Sivapullaiah, P. V 2017 Swelling behaviour of expansive soil treated with fly ash–GGBS based binder. *Geomechanics and Geoengineering*, **12**(3), 191-200.

[10] Prashanth, J. P., Sivapullaiah, P. V., & Sridharan, A 2001 Pozzolanic fly ash as a hydraulic barrier in landfills. *Engineering Geology*, **60**(1-4), 245-252.

[11] Onyelowe, K. C., Shakeri, J., Amini-Khoshalan, H., Usungedo, T. F., & Alimoradi-Jazi, M 2022 Computational Modeling of Desiccation Properties (CW, LS, and VS) of Waste-Based Activated Ash-Treated Black Cotton Soil for Sustainable Subgrade Using Artificial Neural Network, Gray-Wolf, and Moth-Flame Optimization Techniques. *Advances in Materials Science and Engineering*, *2022*.

[12] Armaghani, D. J., & Asteris, P. G 2021 A comparative study of ANN and ANFIS models for the prediction of cement-based mortar materials compressive strength. *Neural Computing and Applications*, **33**(9), 4501-4532.

[13] Nagaraju, T. V., Prasad, C., & Raju, M. J 2020 Prediction of California bearing ratio using particle swarm optimization. In *Soft Computing for Problem Solving* (pp. 795-803). Springer, Singapore.

[14] Nagaraju, T. V., & Prasad, C 2020 New prediction models for compressive strength of GGBS-based geopolymer clays using swarm assisted optimization. In *Advances in Computer Methods and Geomechanics* (pp. 367-379). Springer, Singapore.

[15] Ghorbani, A., & Hasanzadehshooiili, H 2018 Prediction of UCS and CBR of microsilica-lime stabilized sulfate silty sand using ANN and EPR models; application to the deep soil mixing. *Soils and foundations*, **58**(1), 34-49.

[16] Ikeagwuani, C. C., Obeta, I. N., & Agunwamba, J. C 2019 Stabilization of black cotton soil subgrade using sawdust ash and lime. *Soils and Foundations*, **59**(1), 162-175.

[17] Ghiassi, M., Saidane, H., & Zimbra, D. K 2005 A dynamic artificial neural network model for forecasting time series events. *International Journal of Forecasting*, **21**(2), 341-362.

[18] Abiodun, O. I., Jantan, A., Omolara, A. E., Dada, K. V., Mohamed, N. A., & Arshad, H 2018 State-of-the-art in artificial neural network applications: A survey. *Heliyon*, **4**(11), e00938.

[19] Mata, J. (2011). Interpretation of concrete dam behaviour with artificial neural network and multiple linear regression models. *Engineering structures*, *33*(3), 903-910.

[20] Jeremiah, J. J., Abbey, S. J., Booth, C. A., & Kashyap, A 2021 Results of application of artificial neural networks in predicting geo-mechanical properties of stabilized clays—a review. *Geotechnics*, **1**(1), 147-171.

[21] Salehi, M., Bayat, M., Saadat, M., & Nasri, M 2022 Prediction of unconfined compressive strength and California bearing capacity of cement-or lime-pozzolan-stabilised soil admixed with crushed stone waste. *Geomechanics and Geoengineering*, 1-12.

Assessment of Effect of Glass Fiber on the Geotechnical Properties of Black Cotton Soil

Dharmendra Singh[1], Vijay Kumar[2] and R.P. Tiwari[3]

[1] Research Scholar, Department of Civil Engineering, Motilal Nehru National Institute of Technology Allahabad, Prayagraj 211004, India
[2] Assistant Professor, Department of Civil Engineering, Motilal Nehru National Institute of Technology Allahabad, Prayagraj 211004, India
[3] Professor, Department of Civil Engineering, Motilal Nehru National Institute of Technology Allahabad, Prayagraj 211004, India

E-mail: dharmendra@mnnit.ac.in

Abstract. Black cotton soil (BCS) has now been declared as one of the poorest soil among all soils regarding civil engineering construction projects due to presence of its inherent clay mineral i.e. montmorillonite, which introduces undesirable shrinkage and swelling in the BCS. BCS should not be used directly in civil engineering projects because of its excessive volume change behavior. The sample of black cotton soil for the study has been collected from district Narsinghpur near the Bikrampur railway station, Madhya Pradesh, India. Evaluation of effect of glass fiber addition on the strength properties of BCS is described in this research article. Glass fiber acquires good properties such as stiffness, high strength, resistance and flexibility to chemical loss. Glass fiber has been mixed with oven dried BCS from 2% to 8% by its weight. Geotechnical properties are thoroughly investigated before and after treating the BCS sample by means of various laboratory tests. For both treated and untreated BCS samples, UCS testing was performed with varying curing time of 0, 4, 7, 14, 28 and 56 days. UCS value was found to be decreased due to water holding capacity of glass fiber. The laboratory findings indicate that for all additives, MDD and expansion ratio of the BCS decreases, whereas OMC and CBR value increases. XRD (X-ray diffraction) analysis has been conducted on untreated and glass fiber treated BCS samples to estimate change in microstructure and mineralogical composition.
Keywords: Black Cotton Soil; Glass Fiber; UCS; CBR; XRD.

1. Introduction

A lot of areas in arid as well as semiarid regions of the Canada, United States, Argentina, Algeria, Sudan, China, Israel, Spain, India and Australia inherently possess BC Soil [1-2]. BC soil is one of typical expansive soil that spreads over more than 20% of total geographical area of India [3-6]. It covers various regions of Madhya Pradesh, Andhra Pradesh, Gujrat, Maharashtra, Odisha, Tamil Nadu as well as some parts of Uttar Pradesh too [7-8]. BC Soil is found to be one of the most unreliable soil in any construction project since it depicts excessive volume change with varying moisture content [2, 9]. Because these types of soils have a low shear strength and high compressibility, building on weak and soft soils poses a significant risk. which, if improperly handled, can result in significant harm to structures such as runways, spread footings, airport, roads, earth dams and highways [10]. Because clayey soils volume change behavior is so sensitive to moisture content, expansive soils are used difficult in engineering applications [2]. A variety of soil stabilization techniques have been used to stabilize expansive soil, including fly ash [11] RBI Grade-81 [12], Steel fiber [13], Rice husk [14], glass fiber [15-26], bagasse ash [4] [27-28], lime [27, 29-30], cement [26, 31], corn straw ash [32], Polypropylene Fiber [10, 18, 33], Basalt Fiber [34], Coir Fiber [35], Sisal Fibre [36], Jute Fiber [37] and Plastic Waste Powder [38] etc. When the area to be treated is limited, reinforcements are especially trustworthy and successful at strengthening weak soil. Additionally, helping to change the soil's permeability and compressibility is discrete fiber reinforcement. There are many places where fiber-reinforced soil is used, including slopes, retaining walls, and pavements [39]. While traditional geosynthetics like geotextile and Geogrid have been shown to be effective and are being used more frequently in geotechnical engineering, the usage of glass fibers in soil has only recently begun. Despite some research being conducted with a few different types of soil, the volume of research is still quite little when compared to other soil reinforcing materials [15, 40].

1.1. Black Cotton Soil

Sample BC soil was collected from Narsinghpur district near Bikrampur railway station in Madhya Pradesh, India. Coordinates of the soil sample collection site location is reported as 23º07′97″ N and 79º58′74″ E. Figure 1(a) shows image of the soil sample in its natural state. According to IS 1498-1970, extracted soil sample is classified as a CH i.e. inorganic clay with high plasticity. The soil depicts massive swelling and high degree of expansion. Black cotton soil (BC Soil) and Regur soil are local names of expansive soil (ES). Geotechnical properties of collected sample soil are shown in Table 1. Mineralogical elements of collected sample soil are shown in Table 2. FESEM image of virgin soil sample is shown in Fig 2.

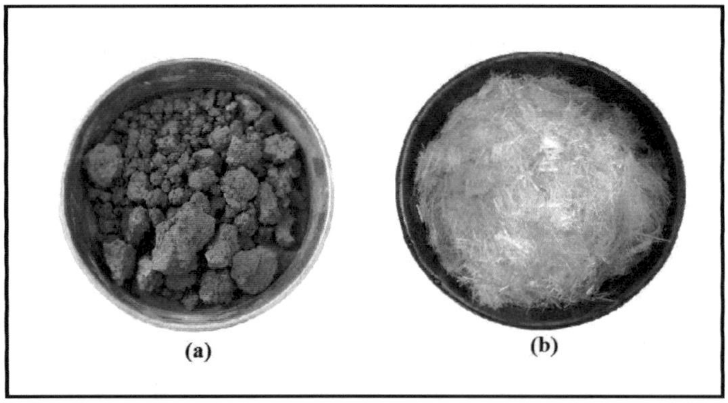

Figure 1. Used materials (a) Black cotton (BC) soil; (b) Glass fiber

Figure 2. BC soil FESEM micrographs

1.2. Glass Fiber

Glass fiber (GF) was purchased from OCV Technical Fabrics Kanpur, Uttar Pradesh, India. Various properties of glass fiber are enlisted below in Table 3. Glass fiber is a lightweight, extremely strong as well as a robust material among all other additives. Nowadays, it is widely adaptable industrial material. Glass Fiber used in this study was procured locally and was available in the form of mesh. Figure 1(b) shows image of the glass fiber used in this study.

Table 1. BCS basic properties

S. No.	Properties		Values
1.	Specific gravity (G)		2.71
2.	Natural moisture content (NMC) (%)		25.91
3.	Free swell index (FSI) (%)		139.86
4.	Liquid limit (LL) (%)		71.80
5.	Plastic limit (PL) (%)		29.07
6.	Plasticity index (PI) (%)		42.73
7.	Shrinkage limit (SL) (%)		10.52
8.	Classification of soil		CH
9.	Compaction	OMC (%)	25.40
		MDD (g/cc)	1.52
10.	UCS (kN/m^2)		139.84
11.	Soaked CBR value (%)		2.45
12.	Gravel (G) fraction (%)		0.00
13.	Sand (S) fraction (%)		1.80
14.	Silt (M) fraction (%)		38.37
15.	Clay (C) fraction (%)		59.83

Table 2. Chemical elements of black cotton soil

S. No.	Element found	Concentration Value (mg/kg)
1	Si	21000.00
2	Fe	4366.67
3	Ca	1343.33
4	K	1053.33
5	Ti	1013.33
6	V	120.33
7	Mn	74.33
8	Sn	26.67
9	Ni	19.00
10	Cr	15.00
11	Zr	15.00
12	Cu	6.67
13	Rb	5.00
14	Zn	3.33
15	Br	1.33

Table 3. Properties of glass fiber from OCV technical fabrics

S. No.	Glass fiber properties		Value
1.	Specific gravity		2.67
2.	Elastic Modulus (GPa)	Tensile	20.2
		Compression	16.9
		Flexural	17.1
3.	Strength (MPa)	Tensile	384
		Compression	201
		Flexural	481
4.	Dimension	Diameter (μm)	21
		Length (mm)	12

2. Methodology and Result Discussion

2.1. Compaction Characteristics

Compaction test was conducted as per IS 2720 (Part-7) 1980 [41] to obtain MDD and OMC of the natural BCS and also same in presence of reinforcing elements. The rammer weighted 2.6 kg was dropped from a height of 310 mm in the compaction test demonstrates soil's water absorption capacity and density too. Cylindrical mould is of 1,000 cm3 capacity, with 100 mm internal diameter and 127.3 mm effective height. Fig.3 shows compaction curve for soils with varying glass fiber content up to 8%. Fig. 4 and 5 show a comparison between compaction curves for both virgin and reinforced soil. It may be clearly observed from the figure that with increasing glass fiber content from 2% to 8%, maximum dry density reduces whereas optimum moisture content increases.

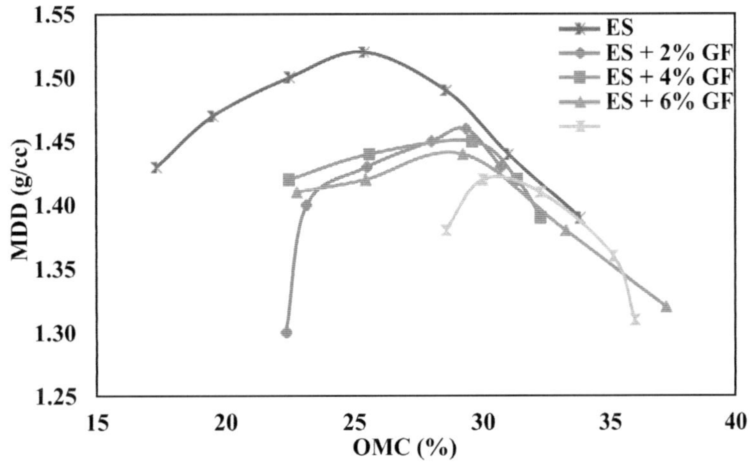

Figure 3. Compaction curve of BCS mixed with GF

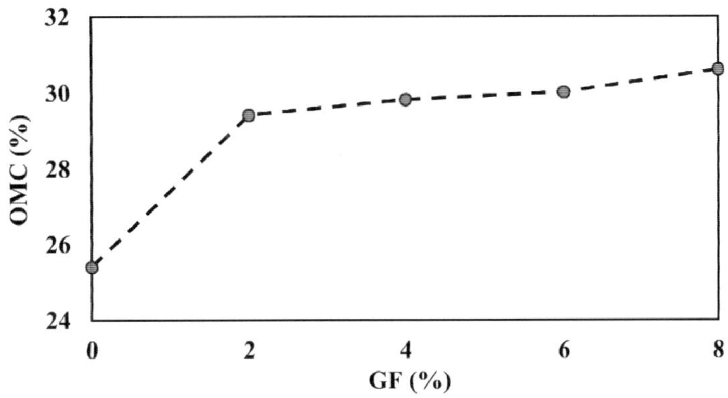

Figure 4. Effect of GF on OMC

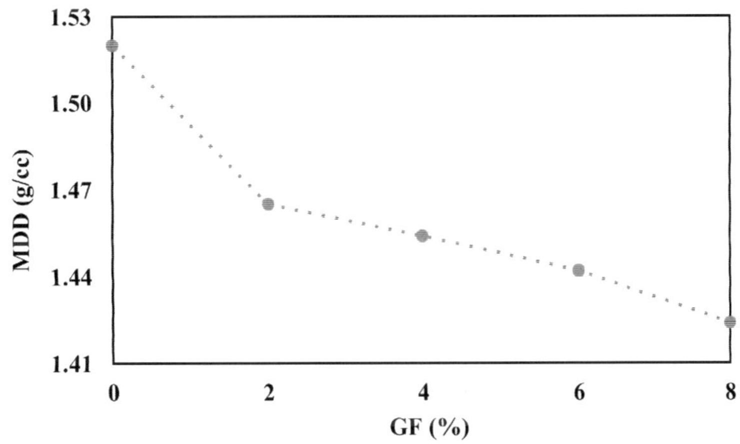

Figure 5. Effect of GF on MDD

2.2. Unconfined Compressive Strength Characteristics

UCS value was determined as per IS:2720 Part-10 (1991) [42]. Specimens were pre-pared for the UCS test with a length-to-diameter ratio of 2 (38 mm diameter and 76 mm length). Cylindrical sample was made with consistent compaction and accurate amount of moisture. Specimens were kept in desiccators till day 0, 4, 7, 14, 28 and 56 after treatment, prior to its testing. 0-day sample has been tested after 4 hours from removing it from desiccator. When samples were tested at day 0, 28 and 56 respectively after curing, the obtained values were found to be decreased as compared to soil without curing. This decrease is due to water holding capacity of glass fiber as shown in Fig. 6 and Fig. 10-11. When samples were tested at day 4, 7 and 14 respectively, UCS values were found to be increased with respect to the soil in its original state as shown in Fig. 7-9. Fig. 12 shows when glass fiber percentages were increased from 0 to 4%, UCS value is maximum. After that UCS value starts decreasing continuously up to 8% of glass fiber content.

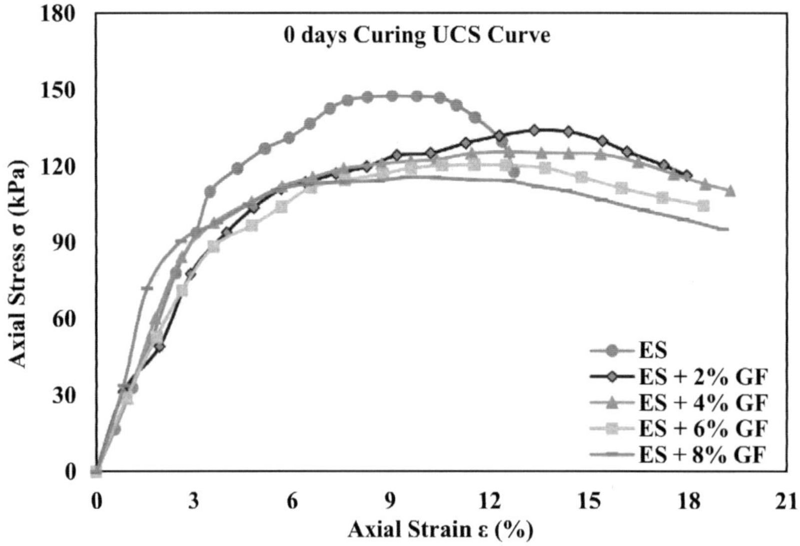

Figure 6. Stress-Strain curve of BCS mixed with GF cured at 0 day

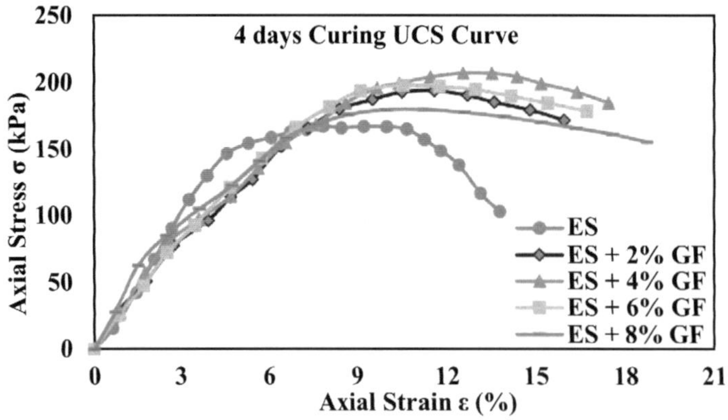

Figure 7. Stress-Strain characteristics of BCS mixed with GF cured at 4 days

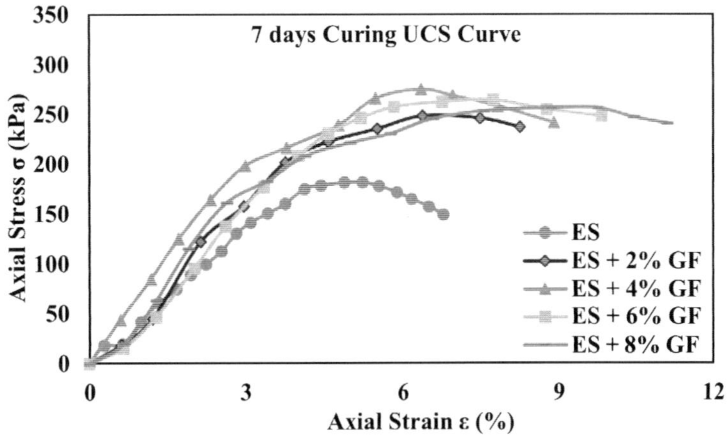

Figure 8. Stress-Strain characteristics of BCS mixed with GF cured at 7 days

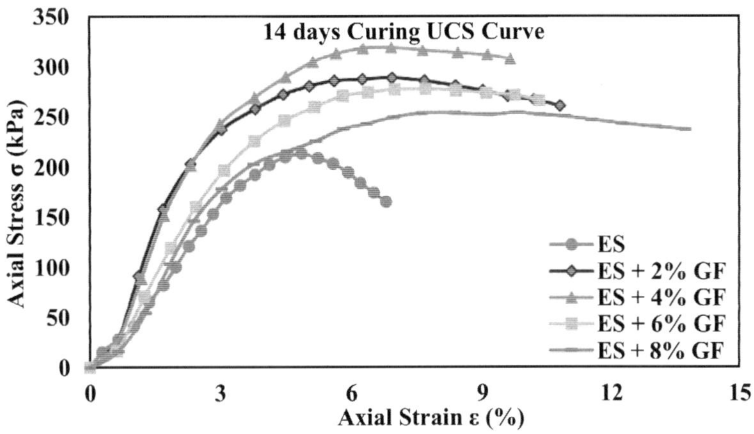

Figure 9. Stress-Strain characteristics of BCS modified with varying Glass Fiber percentages cured at 14 days

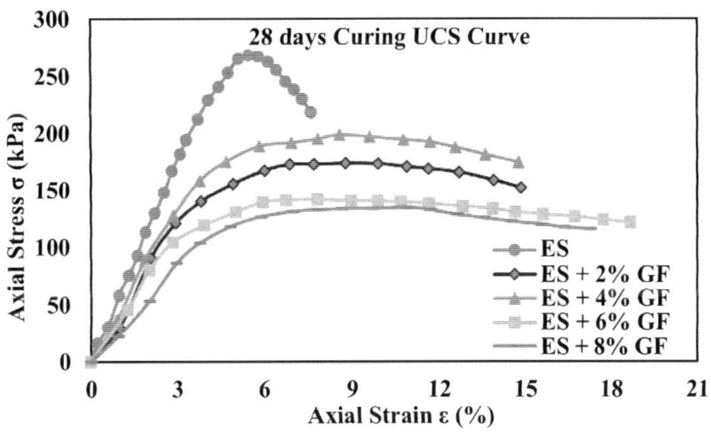

Figure 10. Stress-Strain characteristics of BCS modified with varying Glass Fiber percentages cured at 28 days

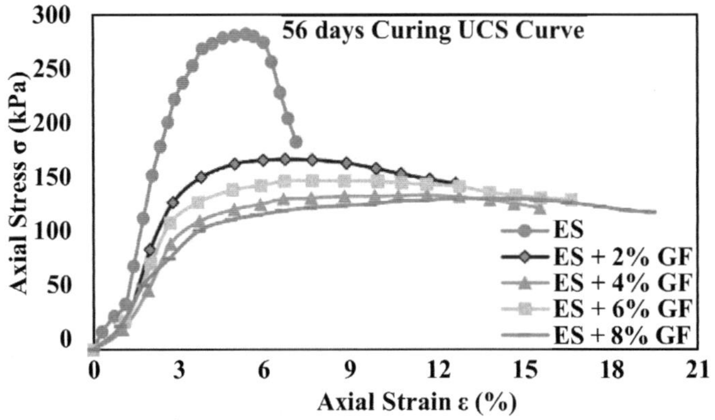

Figure 11. Stress-Strain characteristics of BCS modified with varying Glass Fiber percentages cured at 56 days

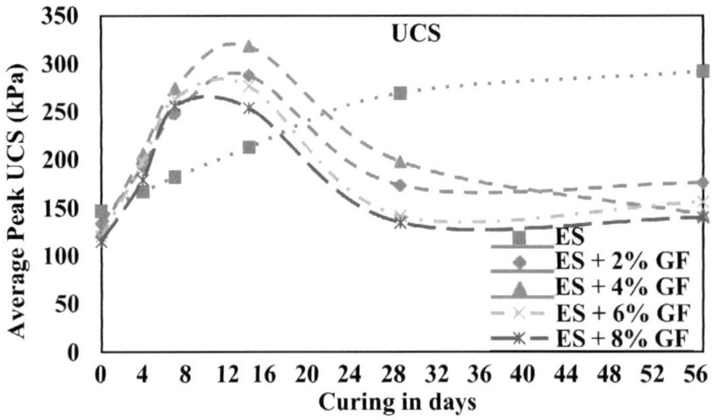

Figure 12. Curing in days with UCS value at 0, 4, 7, 14, 28 and 56 days

2.3. California Bearing Ratio

CBR test has been performed as per IS 2720 (Part-16) 1973 [43]. Test apparatus consists of a 150-mm diameter mould with a base plate and a collar, also a loading frame with a 50-mm diameter cylindrical plunger having 1.25 mm/min penetration rate. 2.5 mm and 5-mm penetrations were induced in soil sample by means of certain load values. Since, CBR value at 2.5 mm penetration is often greater than CBR value at 5 mm penetration, it is usually preferred in pavement design. The test is repeated one more time when CBR value at 5 mm penetration is found to be greater than that at 2.5mm penetration [26]. Fig. 13 shows a load-penetration curve of CBR values of BC soil mixed with 2 to 8% of glass fiber. When 2 to 8% glass fiber is added with BC soil, CBR values were found to be increased continuously as shown in Fig.14.

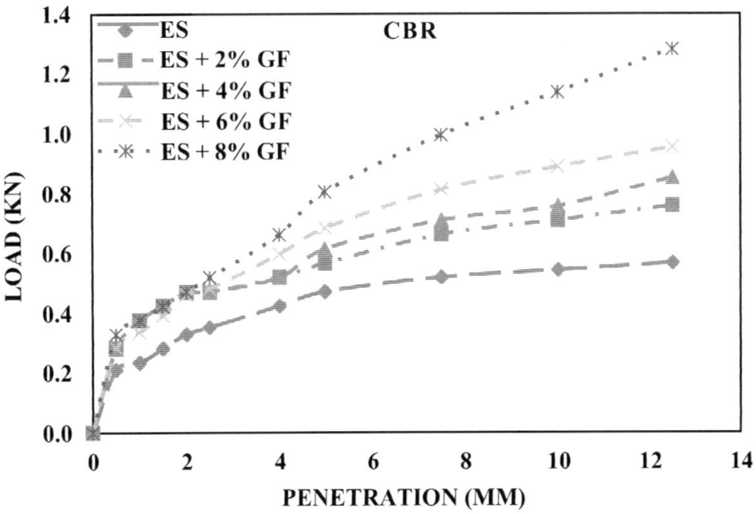

Figure 13. Load- Penetration curve for BCS mixed with GF various percentages

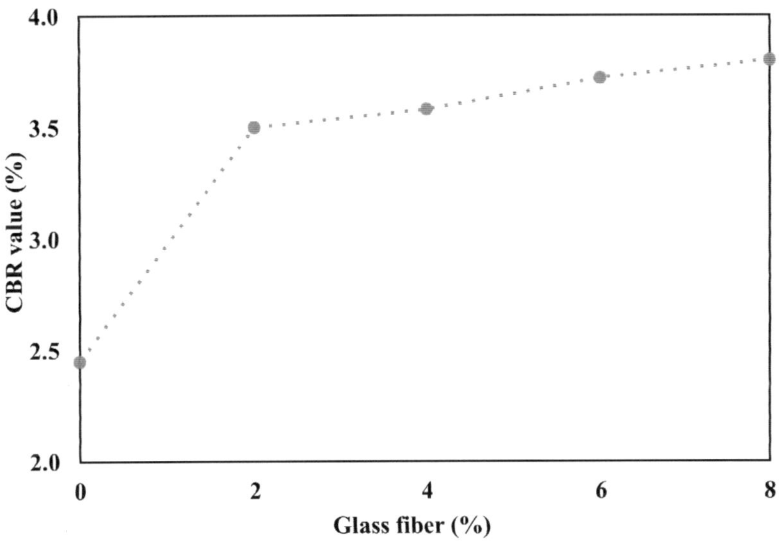

Figure 14. CBR value with various percentages of glass fiber

2.4. Expansion Ratio

Expansion ratio for the sample soil has been determined as per IS 2720 part-16 (1987) [43]. When glass fiber is added with BC soil, its ER was found to be decreased continuously as shown in Fig. 15.

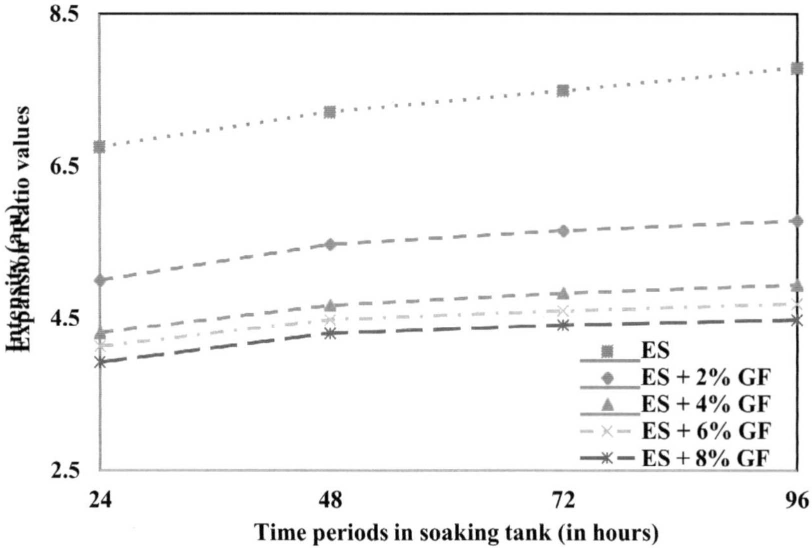

Figure 15. Expansion ratio value with various time periods in soaking tank

2.5. XRD (X-Ray Diffraction)

The XRD patterns in Figure 16-17 reveal that quartz was present in the soil created by black cotton soil and glass fiber. For glass fiber reinforced black cotton soil, a decrease in the strength of quartz peaks indicates that the mineral is consumed during the stabilization process.

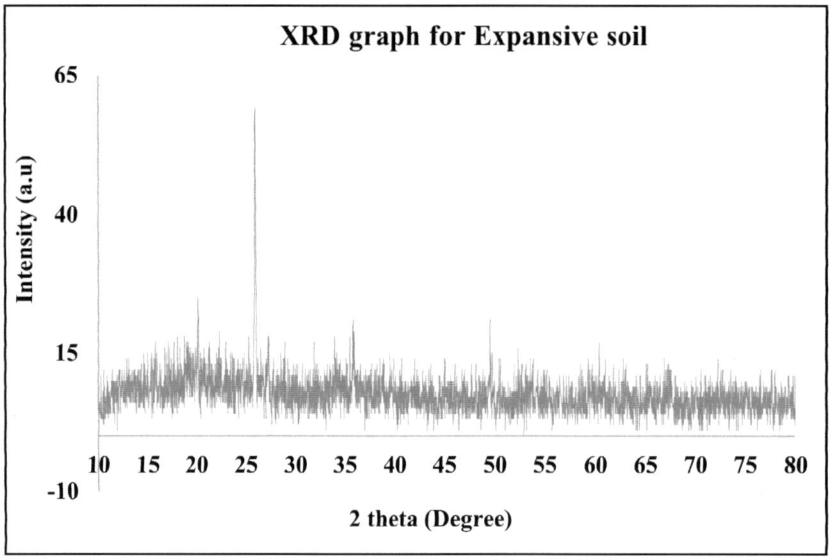

Figure 16. Black cotton soil XRD pattern

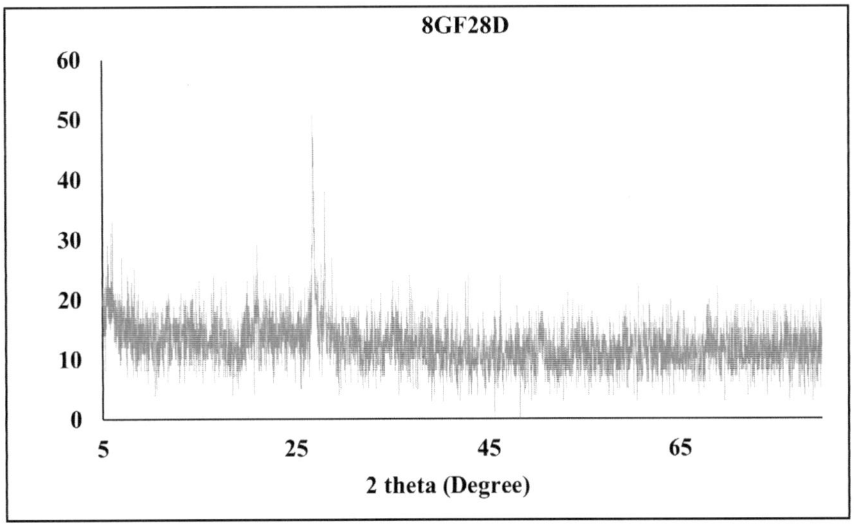

Fig. 17. XRD pattern of Black cotton soil mixed with glass fiber (8%) after 28 days cured

3. Conclusion

Use of glass fiber may slightly enhance the properties of expansive soils. Glass fiber can be used as a modifier with black cotton soil up to a certain limit. The improved geotechnical properties of BC Soil are as follows:

1. MDD, expansion ratio and UCS values were found to be decreased due to entrainment of glass fiber with BC Soil
2. OMC and CBR values were found to be increased with increasing glass fiber percentage from 2 to 8%.
3. With the increase in glass fiber content from 2 to 8% and also curing time, there was a significant decline observed in peak values in the remarkable findings by XRD analysis.

4. References

[1] Fondjo, A.A., Theron, E., Ray, R.P.: Stabilization of Expansive Soils Using Mechanical and Chemical Methods: A Comprehensive Review. Civil Engineering and Architecture, (2021). DOI: 10.13189/cea.2021.090503

[2] Sharo, A.A., Alawneh, A.S., Zghool, H.N., Rabab'ah, S.R.: Effect of Alkali-Resistant Glass Fibers and Cement on the Geotechnical Properties of Highly Expansive Soil. J. Mater. Civ. Eng., 2022, 34(2): 04021417, ISSN 0899-1561, (2022). DOI: 10.1061/(ASCE)MT.1943-5533.0004058

[3] Ramasubbarao, G.V.: Strength behaviour of kerosene coated coir fiber-reinforced expansive soil. Architecture and Civil Engineering Vol. 12, No 2, 2014, pp. 113 – 120, (2014). DOI:10.2298/FUACE1402113R

[4] Kharade, A.S., Suryavanshi, V.V., Gujar, B.S., Deshmukh, R.R.: Waste product 'bagasse ash' from sugar industry can be used as stabilizing material for expansive soils. International Journal of Research in Engineering and Technology, SSN: 2319-1163, Volume: 03 Issue: 03, (2014)

[5] Oza, J.B., Gundaliya, P.J.: Study of black cotton soil characteristics with cement waste dust and lime. Non-circuit branches of the 3rd Nirma university international conference on engineering, Procedia engineering, (2013). doi: 10.1016/j.proeng.2013.01.017

[6] Kumar, S., Sahu, A.K., Naval, S.: Influence of Jute Fibre on CBR Value of Expansive Soil. Civil Engineering Journal, Vol. 6, No. 6, (2020). http://dx.doi.org/10.28991/cej-2020-03091539

[7] Garg, R., Biswas, T., Alam, M.D.D, Kumar, A., Siddharth, A., Singh, D.R.: Stabilization of expansive soil by using industrial waste. Journal of Physics: Conference Series, ICAPSM 2021, (2021). doi:10.1088/1742-6596/2070/1/012238

[8] Ahmed, N. A. K., Damgir, R. M., Hake, S. L.: Effect of fly ash and RBI Grade 81 on black cotton soil as a subgrade for flexible pavements. International journal of innovations in engineering and technology, Volume 4, Issue 1, ISSN 2319-1058, (2014)

[9] Twinkle, S., Sayida, M.K.: Effect of polypropylene fibre and lime admixture on engineering properties of expansive soil. Proceedings of Indian Geotechnical Conference December 15-17, Kochi (Paper No.H-085), (2011)

[10] Soganc, A.S.: The Effect of Polypropylene Fiber in the Stabilization of Expansive Soils. World Academy of Science, Engineering and Technology International Journal of Geological and Environmental Engineering Vol:9, No:8. ISNI:0000000091950263, (2015)

[11] Sabat, A.K., Pradhan, A.: Fiber Reinforced- Fly Ash Stabilized Expansive Soil Mixes as Subgrade Material in Flexible Pavement. Electronic Journal of Geotechnical Engineering, Vol. 19, (2014)

[12] Kodicherla, S. P. K., Nandyala, D. K.: Effect of RBI Grade 81 on strength characteristics of clayey subgrade. International journal of geo-engineering, (2017), https://doi.org/10.1186/s40703-017-0061-z

[13] Irshayyid, E.J., Fattah, M.Y.: The Performance of Shear Strength and Volume Changes of Expansive Soils Utilizing Different Additives", 2nd International Conference on Sustainable Engineering Techniques, Materials Science and Engineering, (2019) doi:10.1088/1757-899X/518/2/022043

[14] Thirumalai, S.R., Babu, V.S., Nirmal, N.R., Lokesh, G.: A Review on Stabilization of Expansive Soil Using Industrial Solid Wastes. Engineering, 9, 1008-1017, ISSN 1947-3931, (2017), DOI: 10.4236/eng.2017.912060.

[15] Baruah, H.: Effect of glass fibers on red soil", International journal of advanced technology in engineering and science, vol. no. 1, special issue no. 1, November 2015, ISSN 2348-7550, (2015)

[16] Shukla, R.P., Algotar, D., Rathod, M.: Effect of Marble Dust and Glass Fiber on Expansive Soil. Journal of the Polish Mineral Engineering Society, (2020) http://doi.org/10.29227/IM-2020-02-50.

[17] Changizi, F., Haddad, A.: Effect of Nanocomposite on the Strength Parameters of Soil. KSCE Journal of Civil Engineering, pISSN 1226-7988, eISSN 1976-3808, (2017), DOI 10.1007/s12205-016-1471-8

[18] Syed, M., Agarwal, S., Guha Ray, A.: Stabilization of Expansive Black Cotton Soil Using Alkali-Activated Binder with Glass and Polypropylene Fiber. Proceedings of the Indian Geotechnical Conference 2019, Lecture Notes in Civil Engineering 136, (2019), https://doi.org/10.1007/978-981-33-6444-8_11

[19] Hashem, M.D., Mahmoud, A.A.H., Ahmed M.A.B., Hag, A.A.: Stabilization of expansive subgrade soil by using additives. Journal of Engineering Sciences Assiut University, Faculty of Engineering, Vol. 44, No. 2, pp. 122 – 131, (2016)

[20] Mukherjee, K., Mishra, A.K.: Hydro-Mechanical Properties of Sand-Bentonite-Glass Fiber Composite for Landfill Application", KSCE Journal of Civil Engineering, pISSN 1226-7988, eISSN 1976-3808, (2019), DOI 10.1007/s12205-019-2015-9

[21] Gul, N., Mir, B.A.: Parametric study of glass fiber reinforced fine grained soil with emphasis on microstructural analysis. International journal of geotechnical engineering 2022, VOL. 16, NO. 6, 716–728, (2022), https://doi.org/10.1080/19386362.2022.2049524

[22] Patel, S.K., Singh, B.: A Comparative Study on Shear Strength and Deformation Behaviour of Clayey and Sandy Soils Reinforced with Glass Fibre. Geotech Geol Eng, 38:4831–4845, (2020), https://doi.org/10.1007/s10706-020-01330-5

[23] Valipour, M., Shourijeh, P.T., Mohammadinia, A.: Application of recycled tire polymer fibers and glass fibers for clay reinforcement. Transportation geotechnics, (2021), https://doi.org/10.1016/j.trgeo.2020.100474

[24] Ddd Bhadoriya, A., Kansal, D.R.: Black Cotton Soil Stabilization using Plastic and Glass Fibres. International Research Journal of Engineering and Technology (IRJET), Volume 05, Issue 11, Nov 2018, e-ISSN: 2395-0056, p-ISSN: 2395-0072, (2018)

[25] Bagriacik, B.: Experimental Study about Soil Improvement with Glass Fibers. International Journal of Engineering Research, Volume No.6, Issue No.8, pp: 392-396, ISSN:2319-6890, (2017), DOI: 10.5958/2319-6890.2017.00045.9

[26] Ruan, B., Zheng, S., Teng, J., Ding, H., Chao M.: Analysis on the Triaxial Shear Behavior and Microstructure of Cement-Stabilized Clay Reinforced with Glass Fibers. Hindawi, Advances in Civil Engineering, Volume 2020, Article ID 8842091, 12 pages, (2020), https://doi.org/10.1155/2020/8842091

[27] Dang, L.C., Hasan, H., Fatahi, B., Jones, R., Khabbaz, H.: Enhancing the engineering properties of expansive soil using bagasse ash and hydrated lime. International Journal of GEOMATE, Vol. 11, Issue 25, pp. 2447-2454, ISSN: 2186-2982, (2016)

[28] Bhujbal, M.P., Gaikwad, M.A.B.: Quantitative study of Shrinkage and Crack Development in Black Cotton Soil. International Conference on Contents, Computing & Communication, (2022).

[29] Wubshet, M., Tadesse, S.: Stabilization of Expansive Soil using Bagasse Ash and Lime", Journal of EEA, Vol. 32, (2014)

[30] Amena, S., Chakeri, D.: AStudy on the Effects of Plastic Waste Strips and Lime on Strength Characteristics of Expansive Soil. Hindawi, Advances in Civil Engineering, Volume 2022, Article ID 6952525, 6 pages, (2022), https://doi.org/10.1155/2022/6952525

[31] Md. M., Mathada, V.S., Moinuddin, M.K., Kotagi, P.M.: Stabilization of Black Cotton Soil using GGBS, Glass Fiber and Ordinary Portland Cement. IJIRST –International Journal for Innovative Research in Science & Technology, Volume 5, Issue 1, ISSN: 2349-6010, (2018)

[32] Audu, M.T., Mbaezue, D.N., Dauda, S.: The engineering characteristics of black cotton soil treated with corn straw ash. International Journal of Advance Engineering and Research Development, Volume 6, Issue 10, e-ISSN: 2348 - 4470, print-ISSN: 2348-6406, (2019)

[33] Priya, C.M.S., Archana, S., Albert, A.B., Deeraj, A.D.: Stabilization of clayey soil using polypropylene fiber. International Research Journal of Engineering and Technology (IRJET), Volume: 04 Issue: 04, ISSN: 2395 -0056, (2017)

[34] Chen, W., Zhao, J., Fan, L., Li, J., Yuan, B., Li, H., Jiang, G., Li, H., Chen, T.: The Effect of Length and Content of Fiber on Glass Fiber and Basalt Fiber-Reinforced Granite Residual Soil. Advances in Civil Engineering, Article ID 7803002, (2022), https://doi.org/10.1155/2022/7803002.

[35] Maurya, S., Sharma, A.K., Jain, P.K., Kumar, R.: Review on Stabilization of Soil Using Coir Fiber. International Journal of Engineering Research, Volume No.4, Issue No.6, pp: 296-299, ISSN:2319-6890, (2015)

[36] Mathew, A., Raneesh. K.Y.: Effect on Strength Characteristics of Expansive Soil Using Sisal Fibre and Waste Materials. International Journal of Science and Research (IJSR), ISSN 2319-7064, (2016)

[37] Wang, Y.X., Guo, P.P., Ren, W.X., Yuan, B.X., Yuan, H.P., Zhao, Y.L., Shan, S.B., Cao, P.: Laboratory Investigation on Strength Characteristics of Expansive Soil Treated with Jute Fiber Reinforcement. International Journal of Geomechanics, ASCE, ISSN 1532-3641, (2017), DOI: 10.1061/(ASCE)GM.1943-5622.0000998

[38] Gowtham S., Naveenkumar A., Ranjithkumar R., Vijayakumar P., Sivaraja M.: Stabilization of Clay Soil by Using Glass and Plastic Waste Powder. International Journal of Engineering and Techniques - Volume 4 Issue 2, ISSN: 2395-1303, (2018)

[39] Sujatha, E.R., Atchaya, P., Darshan, S., Subhashini, S.: Mechanical properties of glass fibre reinforced soil and its application as subgrade reinforcement. Road materials and pavement design 2021, VOL. 22, NO. 10, 2384–2395, (2021), https://doi.org/10.1080/14680629.2020.1746387

[40] Amakye, S.Y., Abbey, S.J.: Understanding the performance of expansive subgrade materials treated with non-traditional stabilizers: A review. Cleaner engineering and technology,(2021), https://doi.org/10.1016/j.clet.2021.100159

[41] IS: 2720 (Part 7)-1980. Methods of test for soils: Part 7 determination of optimum moisture content-maximum dry density relation using light compaction.

[42] IS: 2720 (Part 10)-1973. Methods of test for soils: Part 10 determination of unconfined compressive strength.

[43] IS: 2720 (Part l6)-1979. Methods of test for soils: Part I6 laboratory determination of CBR.

AUTHOR INDEX